SPREADSHEET ANALYSIS FOR ENGINEERS AND SCIENTISTS

SPREADSHEET ANALYSIS FOR ENGINEERS AND SCIENTISTS

S. C. BLOCH

A Wiley-Interscience Publication

JOHN WILEY & SONS, INC.

New York · Chichester · Brisbane · Toronto · Singapore

Library of Congress Cataloging in Publication Data:

Bloch, S. C. (Sylvan Charles)
    Spreadsheet analysis for engineers and scientists / Sylvan C.
Bloch.
        p.   cm.
    "A Wiley-Interscience publication."
    Includes index.
    ISBN 0-471-12683-7 (acid-free)
    1. Signal processing—Data processing.   2. Electronic
spreadsheets.   I. Title.
TK5102.9.B56   1995
621.383'2'0285—dc20   95-10234
                                              CIP

To
Professor Edward Augustine Desloge

Contents

Chapter 4 Reverberation and Q 63

Chapter 5 More About The Decibel and Q 82

Chapter 6   Low-Pass and High-Pass Filters   97

Chapter 7   Band-Pass and Band-Stop Filters 120

Chapter 10   Advanced FFT Operations, II 187

Chapter 11   Signals, Noise, and Decibels 226

Appendix 1   Decibel Definitions   246

Appendix 2   Signal Parameters   253

Appendix 3   Hearing Aid Tests   261

Appendix 4   Wonders of the Universe   266

Appendix 5   Details of Simple Filters   277

Appendix 6   Digits, Bits, and Decibels   294

Index   304

Worksheets

File Name	Graph Names Connected to Worksheet
ABC-OF-Q.WK1 or .XLS	Amplitude vs. t
See Chapter 4 for instructions	dB
Effects and measurement of Q	Power & Phase
	Phase
	Power vs. t
This worksheet lets you experiment with effects of the time constant on Q.	
ADD-DB.WK1	dB SIL or Power
See Chapter 2 for instructions	dB SPL or Volts
The first section is for addition of dB in SPL or voltage. The second section is for addition of dB in SIL or power. Input the dB information and the worksheet calculates the sum, and shows the input and output graphically.	
ANALYTIC.XLS	Analytic Signal
See Chapter 10 for instructions	Spectrum, non-A
Analytic and non-analytic signals in time and frequency	Spectrum, A
	Polar, A
	Power Spec, A
Waveforms and spectra of user-defined analytic and non-analytic signals.	
BANDPASS.WK1	Gain
See Chapter 6 for instructions	Gain dB
Single pole *LCR* band-pass filter	Gain dB [Norm]
(continued)	Gain [Bar]

File Name	Graph Names Connected to Worksheet
BANDPASS (continued)	Gain [Norm]
	Phase
	Phase [Norm]
Input values of inductance L, capacitance C, and resistance R to construct a simple band-pass filter. The worksheet computes the center frequency and the Q, and it graphs the filter gain and phase characteristics in several formats.	
BANDSTOP.WK1	Gain
See Chapter 6 for instructions	Gain dB
Single pole LCR band-stop filter	Gain dB [Bar]
	Phase
	Gain and Phase
	Bode (norm)
	Bode
Input values of L, C, and R. The worksheet computes the center or notch frequency and the Q , and it graphs the filter gain and phase in several formats.	
CEPSTRUM.XLS	Input Signals
See Chapter 10 for instructions	Convolved Sigs
Inverse FFT of log of spectrum	Real Cepstrum
	Lo-t Cepstrum
	Input
	Output
This worksheet computes the cepstrum of convolved signals. This is typically used in identifying multi-path reflections by their time delays. The signals do not have to be known.	
CONVOLVE.XLS	h(t)
See Chapter 10 for instructions	Input Signal
Convolution via FFT and IFFT	In & Out Sigs
(continued)	Input PSD

File Name	Graph Names Connected to Worksheet
CONVOLVE (continued)	Output PSD
	Gain, dB
Convolution is the fundamental linear filtering operation. This worksheet can be used as a module in a signal processing system. The example included in the worksheet is that of a single-pole low-pass filter. Use of the FFT produces "circular convolution." Non-FFT time-domain convolution is provided for comparison.	
CROSSCOR.XLS	Cross-correlate
See Chapter 10 for instructions	Cross Spectrum
Cross-correlation function via FFT and IFFT	Non-FFT Corr
This computes the cross-correlation function, which provides a measure of the similarity of two signals and the time delay between their first points. CROSCOR2.XLS is an associated worksheet with a different type of signal. Non-FFT time-domain cross-correlation is provided for comparison.	
CROSSPEC.XLS	Signals
See Chapter 10 for instructions	Cross Pwr Spec
Cross-spectrum via FFT	Pwr Spec #1
This worksheet computes the cross-spectrum of two signals, which shows what frequencies the signals share in common. The cross-spectrum is also a preliminary computation for other functions, such as cross-correlation and coherence.	
DECONVOL.XLS	Input & Output
See Chapter 10 for instructions	In - Out Error
Deconvolution using FFT and IFFT	
Knowing the impulse response or system function, this worksheet uses the output signal to compute what the input was. It is the "Undo" operation for convolution. A test signal is included in the worksheet to evaluate errors.	
DIGIFILT.WK1	Low-pass
See Chapter 5 for instructions	Hi-pass
(continued)	Low-Pass I/O

File Name	Graph Names Connected to Worksheet
DIGIFLT (continued)	Low-pass I/O
Low-pass and high-pass digital filters	Hi-pass I/O
Input any signal with or without user-defined noise and interference. Input timeconstants for low-pass and high-pass digital filters. View the results as functions of time or frequency, and as an input/output plot. These filters operate in the time domain, without use of FFT and IFFT operations. DIGISWEP.WK1 is an associated worksheet that uses a frequency sweep to check out low-pass and high-pass filters.	
FFTPULSE.XLS	Exact M&P
See Chapter 8 for instructions	Exact Polar
FFT of rectangular pulse	Exact Re&Im
	FFT Polar
	FFT Spectrum
	Lin-Lin M&P
	Power Spectrum
	Pulse s(t)
	Uncentered S(f)
Input any signal and this worksheet computes its spectrum via the FFT in several representations. A rectangular pulse is presently in this worksheet. An exact pulse spectrum is provided for comparison with FFT results.	
FFT-TEST.XLS	Input Signal
See Chapter 8 for instructions	Input PSD
Test of FFT and Inverse FFT operations	Real IFFT
	Input - Output
	Input Bode Plot
This worksheet provides a simple test of the FFT and IFFT in your spreadsheet. Input any signal and this worksheet computes its spectrum and then transforms back to the original data domain. The errors in the FFT and IFFT are plotted. Add log frequency axis to Bode plot if desired.	

File Name	Graph Names Connected to Worksheet
FFT-XMPL.XLS	Impulse Response
See Chapter 8 for instructions	System Function
Examples of the use of FFT and Inverse FFT. (Also see the associated worksheet FREQ-RES.XLS , Chapter 9, for frequency resolution enhancement by use of data windows.)	Exact Re&Im
	Lin-Lin M & P
	Power Spectrum
	Uncentered H(f)
	FFT Polar
	Exact Polar
	Exact Mag(f)
	Exact M & P
This worksheet is a tutorial on FFT and IFFT operations and graphical representations of the results. For comparison, results of an exact Fourier transform are included.	
FREQ-RES.XLS	Input Signal
See Chapter 9 for instructions	dB, B-H Window
Window frequency resolution	dB, Rect Window
	Delta Pwr Spec
This worksheet is convenient for comparison of the operation various windows in terms of their effects on spectra of user-defined signals.	
HI-PASS.WK1	dB & PH [Norm.]
See Chapter 5 for instructions	Gain
Single pole high-pass *RC* filter	Gain dB
	Phase
Input values of *R* and *C*. The worksheet displays the cut-off (corner) frequency and provides graphs of the gain and phase characteristics.	

File Name	Graph Names Connected to Worksheet
IDENTIFY.XLS	Input Spectrum
See Chapter 11 for instructions	Input & Output
System identification using FFT	H(f) Mag & Ph
	Recovered h(t)
Using almost any input signal, even random noise, this worksheet computes the System Function and impulse response. Avoid using a test signal with zeros in its spectrum.	
LOGTABLE.WK1	(No graphs)
See Chapter 1 for instructions	
This worksheet provides a table of logs, base 10, to provide a global perspective of the numerical data. There are two sections of this worksheet. The first is a mini-table and the second is more detailed.	
LOW-PASS.WK1	dB & PH [Norm.]
See Chapter 5 for instructions	Gain
Single pole low-pass *RC* filter	Gain [dB]
	Phase
Input values of *R* and *C*. The worksheet displays the cut-off (corner) frequency and provides graphs of the gain and phase characteristics.	
MODE-D-K.WK1	Damped
See Chapter 4 for instructions	dB
Multi-mode decay with different time constants for each mode	Amplitude^2
	Undamped
Input amplitude, frequency, and time constant for two modes. The worksheet displays amplitudes and energy as functions of time, at one point in space, for undamped and damped waves. It also displays envelopes of exponential decays for each time constant.	

File Name	Graph Names Connected to Worksheet
MODEMOVE.WK1	Displace vs X
See Chapter 4 for instructions	Spectrum
View incoherent modes and their sum at random times	Spectrum dB
This is like WAVEMODE except the modes are incoherent. You can view the modes and their sum at random times. Watch the wave group bounce off the walls as you repeatedly press the Recalculate Key (usually Function Key F9).	
PHASE-RA.XLS	dB
See Chapter 10 for instructions	dB FFT
Linear Phased Arrays	FFT Polar
	Norm FFT Pwr
	Power
	Polar
	Beam Width
This worksheet contains two sections. The Home screen section computes a phased array pattern by non-FFT methods, with user-defined array parameters. The second section (press PgDn) computes a phased array pattern by FFT methods, which includes the possiblity of weighting individual elements.	
Q-OF-LCR.WK1 or .XLS	Current: Freq
See Chapter 4 for instructions	Impedance: Freq
Single pole *LCR*	Waveform: Time
	Power: Freq
	Phase: Freq
	dB: Freq
Input values of *L*, *C*, and three values of *R*. This worksheet shows dB, current and impedance in the frequency domain and the waveform for free decay in the time domain for the three damping constants. It also computes the *Q*s and the frequency shifts from the undamped frequency.	

File Name	Graph Names Connected to Worksheet
RT-60.WK1	RT60 [dB]
See Chapter 4 for instructions	Comparison
Reverberation time measurement, to −60 dB for single mode. Press PgDn for a numerical comparison of the Sabine and Norris-Eyring equations showing reverberation time as a function of average absorption coefficient.	
SIGNOISE.WK1	Amplitudes, S&N
See Chapter 11 for instructions	Power, S&N
Signal/Noise power ratio	S+N Amplitude
	Rate vs S/N
	Rate vs BW
This worksheet calculates the signal/noise power ratio as a decimal value and in dB.	
SNR-MSC.XLS	Signals
See Chapter 10 for instructions	Cross Pwr Spec
SNR function and magnitude squared coherence (MSC) function via the FFT	Windowed Signals
	Coherence Funct
	SNR Function
	SNR Function
	Windowed Signals
	Pwr Spc, Sig #1
	Pwr Spc, Sig #2
	dB, Sig #1
	dB, Sig #2
This worksheet computes the SNR function which gives the signal/noise ratio at each frequency. It also computes the MSC (magnitude squared coherence function) which tells how two signals are related at each frequency.	

File Name	Graph Names Connected to Worksheet
SUB-DB.WK1	dB SIL or Power
See Chapter 2 for instructions	dB SPL or Volts

Decibel subtraction. This worksheet contains two sections. One is for subtraction of dB in sound pressure level or voltage. The second subtracts dB in sound intensity level or power. Input the dB information and the worksheet calculates the difference, and shows the input and output graphically.

SYNSPECT.WK1	Amplitude
See Chapter 8 for instructions	Spectrum
Fourier Series	Spectrum 2-D
	Waveform

This worksheet synthesizes a periodic waveform using Fourier series. You enter harmonic number and the Fourier coefficients for the sine and cosine terms, and select the waveform period, time increment, and time start. Graphs show the synthesized waveform using 9 harmonics, and the spectrum amplitude and dB. The worksheet is easily expanded to utilize more harmonics. User-defined random noise can be added to one or more harmonics.

SYS-FUN.XLS	h(t)
See Chapter 10 for instructions	h(t) error
System function using FFT	Imaginary Error
	Mag & Phase

This worksheet shows use of the FFT in computing the System Function $H(f)$ from the impulse response $h(t)$, and use of the IFFT in computing $h(t)$ from $H(f)$.

THERMAL.WK1 or XLS	Thermal Volts
See Chapter 11 for instructions	Blackbody
Thermal noise and radiation	Log Blackbody

The Home screen computes open-circuit thermal noise voltage across a resistor. You can input the value of the resistance and the bandwidth. The worksheet calculates and graphs the thermal noise voltage as a function of absolute temperature. The second screen computes blackbody radiation energy density for inputs of absolute temperatures.

WAVEMODE.WK1	Displace vs X
See Chapter 4 for instructions	Pressure vs X
Coherent standing wave modes in one dimension	Spectrum
	Spectrum dB

Input the amplitude and frequency; the results are shown graphically and numerically, for displacement and pressure as functions of position. This worksheet is set up for three modes but you can easily add additional modes. Each mode and the sum of the modes are shown at one instant in time. Use the worksheet MODEMOVE to view incoherent modes at random times.

WINDOWME.XLS	Input Selected
See Chapter 10 for instructions	Input, Rect Win
Comparison of window results. Common windows are selectable. User-defined windows may be added to the Window Library	Pwr Spec, Rect
	Pwr Spec, Selec
	Pwr Spec, 2 Win
	Delta Pwr Spec

This worksheet contains several types of data windows and performs tests so you can compare the trade-offs in applying them. You can add more windows.

Worksheet Tips

Installing the Diskette Files

To install the files included on the disk, do the following:

1. Assuming you are using drive A: as the floppy drive for your diskette, at the A:> prompt type INSTALL and press Enter↵ .
2. You may also type A: INSTALL and press Enter↵ at the C:> prompt.
3. Follow the instructions displayed by the installation program. The default directory path is C:\SPRDSHT but you may enter a different selection. The spreadsheet files will be installed beneath the main directory in chapter-based subdirectories such as CHAP1. The standard installation option will load Lotus *1-2-3* WK1 files for all of the worksheets (except CROSCOR2 in CHAP10). WK1 files can be imported into any spreadsheet. This option requires about 1.9 MB on your hard disk. If you wish to use the advanced FFT analysis tools available in Microsoft *Excel* 5.0 and later versions you can also toggle on the second option with the space bar. This option installs XLS files for the spreadsheets in Chapters 7, 8, 9, and 10 and requires

an additional 2.2 MB of disk space. When you open a chapter file in your spreadsheet be sure to specify which file type you want, or "all files."

If you have any questions about installing call Wiley Technical Support Services at (212) 850-6194.

Graph Formats

The graphs in *Lotus* WK1 format are set up to run on the simplest spreadsheet. You may want to modify the colors for graphical clarity. All graphs are set up using the left-hand y-axis. The WK1 format does not support graphs with right-hand y-axes. You may want to modify the graphs to use both left and right y-axes if you use recent versions of spreadsheets. This is important, for example, in graphs of magnitude and phase where the phase has a range of $\pm180°$ and the magnitude may only go from 0 to 1. Tip: For better viewing change the phase from degrees to radians. The WK1 format does not support logarithmic axis scales, so you may want to modify some graphs if your spreadsheet has log scaling. To view graphs use standard commands /Graph Name Use. Select the graph name and press Enter↵. Press Function Key F10 to view the current (most recently viewed) graph.

The XLS files are ready to run in *Excel* 5.0 and later versions; to access graphs use the mouse pointer and click on the graph name at the bottom of the screen. Graphs are set up using log scaling and left and right axes where appropriate.

FFT and Complex Functions

Worksheets for Chapters 8, 9, and 10 use complex functions and the FFT operation. The WK1 format does not support complex functions, so frequency domain worksheets using complex functions will require modification to include these functions in the appropriate cells, as described in the Worksheet Organization sections of Chapter 8, 9, and 10. If your spreadsheet does not have complex functions you can still see how they work, but you can't make changes. In *Excel* these functions are called complex engineering functions and the @-symbol is not used. The *Excel* XLS files are ready to run and do not require modification.

Novell/WordPerfect *Quattro Pro* can import WK1 files and it has the FFT and complex @-functions. If your spreadsheet does not have the FFT operation you can use an add-in for this purpose (see Digging Deeper, Chapter 8).

Graphical Zoom

Some of the worksheets let you enter the time or frequency increment as an option. You can use this as a graphical zoom feature by setting the start of the axis and adjusting the increment. Otherwise, use the Fill command to set the start and the increment.

On-Screen Digital Readouts

Some worksheets show you a useful technique using the IF function and the MAX function to obtain digital readouts of time delay, amplitude, maximum crosscorrelation, input and output variances, etc. The IF function is like an exclusive-or gate used in digital logic circuits. This logical function is one of the most powerful, and under-appreciated, functions in spreadsheets. Consult your manual or on-screen Help for details.

Memory Management Tips

Resident Programs and Drivers

The worksheets are not large (less than 200 KB) but if you have Terminate and Stay Resident (TSR) programs on your computer you may run out of memory. Disable unnecessary TSR's and device drivers to make more memory available.

DOS 6 and Above, and Windows

The worksheets on the diskettes are small enough to run on a computer with limited memory. However, if you expand a worksheet to handle a large data set you may run short of memory. If you operate in DOS you can free up more memory by using one of the latest versions.

The EMM386 component of MS-DOS allows programs to access and use up to 32 megabytes beyond the 640K limit of conventional RAM. Other programs such as Quarterdeck Office Systems' *QEMM 386* and Qualitas Inc.'s *386MAX* perform similar functions. Microsoft *Windows* has a 386 control program with basic features.

Preface

dB or not dB? That is the question.

Electronic spreadsheets, originally developed for financial and business applications, have achieved universal popularity with engineers and scientists because of their ease of use, powerful mathematical and statistical capabilities, and excellent graphics. New versions of spreadsheets contain the Fast Fourier Transform, complex operations, and even Bessel functions. Most data acquisition packages provide seamless data transfer to and from spreadsheets. The Dynamic Data Exchange (DDE) feature of the *Windows* interface makes your spreadsheet an integral part of both real-time and off-line data collection and analysis in the laboratory and by modem from remote sources. Via cellular phones, fiber optics, satellite links, and the Internet your spreadsheet is a window to the world, and beyond.

An outstanding feature of spreadsheets is the fast learning curve of the user. Beginners progress rapidly because there is no programming language to master. Everything is under the control of the user and nothing is hidden; all formulas are visible.

The worksheets provided on the diskettes contain no macros so they will work in any spreadsheet. They have been tested in *Excel*, *Quattro Pro*, and *Lotus 1-2-3*.

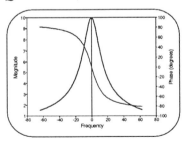

Most of the graphics in this book are produced directly from the worksheets. Of course, the appearance of the graphics on your monitor will depend on what type of spreadsheet you use. Graphs in this book that are similar to ones that you can view on your monitor are shown with a rounded rectangular outline. Most of the graphics are best when viewed in color. The graphics associated with each worksheet are "live" and allow you to view and print results immediately.

There are several excellent books on spreadsheets for engineers and scientists. They discuss interfacing spreadsheets directly to electronic instruments, statistics and curve-fitting, chemical laboratory applications, and signal processing. I have attempted here to provide a wide variety of new applications (of my own choosing). Some of the worksheets here are tutorial to show you some tips and tricks to help you compose your own worksheets for as yet undreamed-of applications. In the long run these may be the most useful to you. Other worksheets can be applied directly to incoming data or design and simulation applications.

This book is interactive. It works best when you are a member of the crew, not just a passenger. Worksheets on diskette are included so you can use your computer to make changes. You can see and print out the results instantly. Don't be afraid to ask "What if . . .?" I hope you will experiment and gain an intuitive and quantitative feeling for the varied topics offered here for your enjoyment. Self-tests are provided so you can check your skills as they progress. No one but you will ever know the results!

In many books the author speaks directly to the reader in the Preface. After this the author is never heard from again. I have taken an informal approach and I hope that the conversational style will be user-friendly. All you need is a little algebra to understand the subject matter here, and even that is reviewed for your convenience.

I am indebted to Professor Silvanus Phillips Thompson for his guiding dictum "What one fool can do, another can." This has been a source of inspiration throughout my life.

I would like to express my appreciation to the editors and staff at John Wiley & Sons for their help and their patience. In alphabetical order, special thanks go to Jack Drucker, Robb Frederick, Perry King, Rose Leo Kish, George J. Telecki, and Lisa Van Horn. I would also like to thank Alan Conrad of *Microwaves & RF* for suggesting the phased array, and Robert Dressler for his many suggestions and insights, too numerous to list.

Tampa, Florida   Sylvan Charles Bloch

Chapter 1

The Adventure Begins

It was a dark and stormy night, and the decibels were howling ominously through the withering heather, still desperately clinging to the surrounding sodden Scottish moors which seemed to crouch, ready to pounce upon and ingest . . . nay, *devour* Merchiston Castle. Inside, on this midnight dreary, while he pondered weak and weary over many a tome of long-forgotten lore, John Napier, the Laird of Merchiston, saw the image of a light bulb suddenly appear above his head. He thought this was more than passing strange because it was the beginning of the seventeenth century and light bulbs had not been invented. Then he realized it was no light bulb, it was the symbol for an idea! More about this later, with the rest of the news, weather, and sports at eleven.

Our story starts with the decibel. Contrary to popular misconceptions, the decibel is not a ten-headed monster. Actually, the decibel is a six-headed monster (see Fig. 1-1).

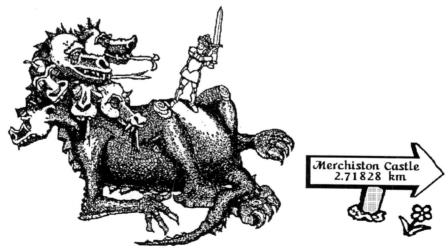

Fig. 1-1. Ordinary mortal (one head) conquers decibel (six heads).

Any ordinary mortal armed with enough courage, armor, and a sword can conquer the decibel. Decibels are just special logarithms, and logarithms are just exponents, so we have to start with exponents. Please fasten your seat belts.

Powers of Ten: Throw Another Logarithm on the Fire

Once upon a time, long ago and far away in a land called Scotland, there lived a wily Scot named John Napier (1550 – 1617) who conceived the idea of logarithms (it wasn't a light bulb, it was a logarithm). Actually, the moors and glens of Scotland were full of logarithms but, like the heather, no one had found a use for them. Napier's simple discovery revolutionized computation and enabled navigators not only to explore the world but also to find their way back home and relate their wonderful adventures.

Scotland became the world's largest exporter of logarithms. Napier, the Laird of Merchiston, was able to reduce multiplication and division to simple addition and subtraction, operations which many captains of ships could perform, given enough time. Napier's discovery also reduced to mere multiplication and division the more complicated operations of raising numbers to powers and taking roots of numbers. Thus modern navigation was born, worldwide trade began, and the blessings of science, technology, philosophy, and the fine arts came to all of humanity.

In 1614 Napier shared his discovery with the entire world in his book *Mirifici Logarithmorum Canonis Descriptio*. As a result crime, war, disease, poverty, dishonesty, and famine were eliminated and everyone lived happily ever after.

An eminent mathematician of the day, Professor Henry Briggs of Oxford (1556 – 1631), was so taken with Napier's logarithms that he made a journey all the way to Scotland to pay homage. In what must be one of the greatest put-downs of all time, the eminent mathematician said, "After hearing of your discovery, I was amazed that anyone could think of such a thing. But, after meeting you, I'm amazed that no one thought of it before." (or, words to that effect). The Laird of Merchiston also invented the decimal point, without which we would all be lost.

Multiplication

Examine the pattern of the following multiplications:

$$10 \times 100 = 1\,000$$

$$100 \times 1\,000 = 100\,000$$

$$1\,000 \times 10\,000 = 1\,0\,000\,000$$

Everyone knows that Scotsmen are famous for their thriftiness. Napier found that he could save a lot of printer's ink by not using so many zeros if he placed a superscript at the upper right hand corner of 10, and the value of the superscript would indicate how many zeros there should be.

A superscript looks like this: 10^{12}. The superscript is 12.

A superscript should not be confused with a script for a superstar. In order to avoid confusion, from now on we will call these superscripts *exponents*. The exponent indicates the power to which the base (10) is raised. Sometimes, especially in computer printouts, you see exponents written like E5. For example, 10^7 would be written as E7, or E+7. This avoids superscripts.

In fact, use of exponents is much more clear than counting zeros, because the zeros are already counted for you:

$$10^1 = 10$$
$$10^2 = 100$$
$$10^3 = 1\ 000$$
$$10^4 = 10\ 000$$
$$10^{16} = 10\ 000\ 000\ 000\ 000\ 000$$

To be consistent we should also include the special case of no zeros: $10^0 = 1$. It will soon become clear why 10 (or any number) with a zero exponent is always equal to 1. Now our pattern of multiplication takes the form shown in Table 1-1.

Table 1-1. Multiplication	
Powers of Ten	Exponents
$10^1 \times 10^1 = 10^2$	$1 + 1 = 2$
$10^1 \times 10^2 = 10^3$	$1 + 2 = 3$
$10^3 \times 10^3 = 10^6$	$3 + 3 = 6$

The pattern is clear: To multiply we add exponents. Check yourself to make certain that you have the above information stored correctly between your ears. The answers are at the end of this chapter.

Self-Test 1-1	
(a) $10^3 \times 10^5 =$	(d) $10^{15} \times 10^3 =$
(b) $10^7 \times 10^8 =$	(e) $10^6 \times 10^9 =$
(c) $10^2 \times 10^9 =$	(f) $10^4 \times 10^6 =$

Divide and Conquer

Now let's see how the rules of exponents apply to division. Up to now we have only used positive exponents. What is the meaning of something like 10^{-7}? Let's perform an experiment:

$$1\ 000/100 = 10^3/10^2 = 10^1$$

In terms of exponents we can express this division as $3 - 2 = 1$.

Let's contemplate division, exponents, and the Meaning of Life. Study the pattern in Table 1-2.

Table 1-2. Division	
Powers of Ten	*Exponents*
$10^1 / 10^0 = 10^1$	$1 - 0 = 1$
$10^3 / 10^1 = 10^2$	$3 - 1 = 2$
$10^2 / 10^3 = 10^{-1}$	$2 - 3 = -1$

Again, the pattern is clear: To divide, we subtract exponents. But now something new has appeared: The meaning of negative exponents! (The Meaning of Life comes later. Trust me.) Now it's your turn to try some division in Self- Test 1-2.

Self-Test 1-2	
(a) $10^6/10^1 =$	(d) $10^3/10^2 =$
(b) $10^5/10^2 =$	(e) $10^4/10^2 =$
(c) $10^8/10^9 =$	(f) $10^7/10^3 =$

The Reciprocal

By the reciprocal of a number, N, we mean $1/N$ or N^{-1}. Notice that division is just a form of multiplication by the reciprocal. For examples, see Table 1-3.

Table 1-3. Reciprocals			
Number	*Reciprocal*	*Fraction*	*Decimal*
0.1	10^1	$10/1$	10.0
10.0	10^{-1}	$1/10$	0.1
100.0	10^2	$1/100$	0.01

Now you know all about some basic operations with exponents, so try your hand at Self-Test 1-3.

Self-Test 1-3	
(a) 10^{-4} = (fraction) = (decimal)	(d) $10^3 / 10^{-2}$ =
(b) 10^{-6} = (fraction) = (decimal)	(e) $10^{-2} / 10^{-4}$ =
(c) $10^{-1}/10^3$ =	(f) $10^5 / 10^7$ =

Logarithms

These exponents that you have been using are called logarithms, to the base ten, or simply logs, base ten. Instead of writing all this out people usually use a subscript notation,

$$\text{logarithm the base } 10 = \log_{10} = \log$$

The $\log_{10}$ is called the common log; this is not meant to be derogatory. It is so named because the people who invented counting had ten fingers. If their circle of awareness had included their feet we would probably now have $\log_{20}$ for common logs. The only other kind of logarithm that we will be using is the natural logarithm, which has the base called "e" (e = 2.71828..., a never-ending, never-repeating, fundamental number of the universe. It's called a transcendental number.) This is one of the Inscrutable Wonders of the Universe, but don't worry about it now (see Appendix 4). The usual notation for the natural logarithm is,

$$\text{logarithm to the base } e = \log_e = \ln$$

From now on, we will drop the subscript on the common log, so when we write "log" we will mean "$\log_{10}$" for common logs and when we write "ln" we will mean "$\log_e$" for natural logs.

General Warning #1: In the BASIC computer programming language the command LOG(x) returns the natural log of x. BASIC does not have common logs. You can get into a lot of trouble if you forget this Fact of Life. Later, if you survive to Chapter 2, we will see how to take care of this by changing bases.

General Warning #2: Never, *never*, NEVER, *NEVER* TRY TO TAKE THE LOGARITHM OF A NEGATIVE NUMBER. Technically speaking, this is a No-No; it is an undefined operation and neither the author nor the publisher can be held legally responsible if you attempt to violate this warning. Without exception, absolutely everyone who has tried this has died or will die.

What you have learned is, to multiply two numbers, add their logarithms. Then find the number which is ten raised to that logarithm; this number is called the antilogarithm, or simply the antilog.

Find The Logarithm

Now you should be able to answer the riddle, What is x?:

$$10^x = 1000$$

Right! Clearly, $x = 3$. So, the log to the base ten, of 1 000, is 3. Also, the antilog of 3 is 1 000. Later we will see how we can find the log of any number.

Raising To A Power

Raising a number to a power is a common operation. We have been doing this before but now we need to generalize our notions and notations. Consider Table 1-4.

Table 1-4. Raising to a Power	
Numbers	Powers of Ten
10 × 10 = 100	$10^1 \times 10^1 = 10^2$
10 × 10 × 10 = 1 000	$10^1 \times 10^1 \times 10^1 = 10^3$

We could write these expressions equally well (or better) as shown in Table 1-5.

Table 1-5. Raising to a Power and Exponents	
Power Notation	Exponents
$(10^1)^2 = 10^2$	$1 \times 2 = 2$
$(10^1)^3 = 10^3$	$1 \times 3 = 3$

In a similar fashion, $100 \times 100 \times 100 = (10^2)^3 = 10^6$. So, to raise a number to a power, multiply the exponent by the power. Now try your hand at Self-Test 1-4.

Self-Test 1-4	
(a) $10^x = 10\ 000$; what is x?	(d) $(10^4)^3 =$ (10 with exponent)
(b) $10^x = 0.001$; what is x?	(e) $(10^2)^3 =$ (10 with exponent)
(c) $(10^2)^2 =$	(f) $(10^{-2})^5 =$ (10 with exponent)

Roots, Square And Otherwise

Music sounded funny before Napier's discovery. That's because musicians needed the twelfth root of 2 (what number, multiplied by itself 12 times, equals 2?) to tune their instruments but the mathematicians of the time could only guess at an approximate answer. Some people think music still sounds funny. Very soon we'll see how to find the twelfth root of 2 and many other exotic things (see Example on page 14), so stay tuned. Extracting roots is automatically included in our system. It merely consists of raising a number to the reciprocal power, as in Table 1-6.

Table 1-6. Roots		
Numbers	*Power Notation*	*Exponents*
$100^{1/2} = 10$	$(10^2)^{1/2} = 10$	$2 \times 1/2 = 1$
$1\,000^{1/3} = 10$	$(10^3)^{1/3} = 10$	$3 \times 1/3 = 1$
$10\,000^{1/2} = 100$	$(10^4)^{1/2} = 10^2$	$4 \times 1/2 = 2$

The first and third examples are called square roots; the second example is called the cube root. In general, the nth root (pronounced *enth* root) of a number is obtained by raising the number to the $1/n$ power. So, to take the nth root of a number, multiply its logarithm (or exponent) by $1/n$.

If Alex Haley could find his Roots, so can you. Try Self-Test 1-5.

Self-Test 1-5	
(a) $1\,000\,000^{1/2} =$	(d) $100^{-1/2} =$
(b) $(10^8)^{1/2} =$	(e) $(10^{-16})^{1/2} =$
(c) $(10^4)^{1/4} =$	(f) $(10^{66})^{1/3} =$

Scientific Notation

It must be evident from what we have been doing that writing numbers as powers of 10 is a convenient, compact, and clear way to express very large and very small numbers, as well as those in between.

When working with almost any kind of numbers, and especially with decibels, you will find it very helpful if you would always write numbers in scientific notation. By scientific notation we mean a number between 1 and 10, times a power of 10. Let's summarize all of this in Table 1-7.

	Table 1-7	
Number	*Power of Ten*	*Exponent*
1000.000	10^3	3
100.000	10^2	2
10.000	10^1	1
1.000	10^0	0
0.100	10^{-1}	−1
0.010	10^{-2}	−2
0.001	10^{-3}	−3

Consider the following examples:

$$2.5 \times 10^2 = 250$$
$$3.75 \times 10^5 = 375\ 000$$
$$7.808 \times 10^{-2} = 0.07808$$
$$1.981 \times 10^3 = 1\ 981$$

There are many advantages to this notation; we can easily compare significant figures and powers of ten, and scientific notation will make it much easier to work with dBs, which is the whole purpose of this, anyway.

Note on Notation: In these troubled times when concern is growing about the squandering of non-renewable resources one must pause and reflect on the vast numbers of exponents required by industrialized nations. This concern is expressed eloquently by APE (Association to Protect the Exponent), whose international headquarters are in La Jolla, California. As an alternative to the superscript notation for exponents, the notation used in computer print-outs is becoming more popular (see p. 3). This notation uses E(x) for powers of 10 and exp[x] for powers of the base of natural logs, thus saving countless exponents for posterity. Consider the following examples:

$10^{5.3} = $ E+5.3	$9.61 \times 10^{-7} = 9.61\text{E}{-7}$
$e^{8.1} = \exp[8.1]$	$3.2 \times e^{-4} = 3.2\exp[-4]$

Computers, typists, and printers are grateful for this notation because all of the symbols are placed on the same line. Wouldn't it be super if this notation would supersede the superscript? In the meantime during this transition period you must be alert to both notations.

Change the following items in Self-Test 1-6 to scientific notation, or change from scientific notation to ordinary positional notation.

Self-Test 1-6	
(a) 125 =	(d) 0.000142 =
(b) 0.455 =	(e) 2.67×10^4 =
(c) 1 678 000 =	(f) 3.555×10^3 =

Putting It All Together

Okay, rehearsal is over and now we're ready to put it all together. We are going to collect all of the ideas above into a neat format so that operations with logarithms can be done in an orderly manner. We will be using a table of common logs; of course in daily life you will probably use a calculator or computer (unless you are stranded on a desert island) but you need to understand what the calculator is doing. Also, you need to tell the calculator what to do and this calls for understanding, not rote memory.

You may feel like hitting the fast-forward button and going directly ahead to Chapter 2, ABCs of dBs. Of course, you can do this, but if you do you will miss the basic knowledge of what's happening and you probably won't get an intuitive appreciation for decibel measurements. Resist the impulse and stay tuned.

Look back at Problem (a) in Self-Test 1-1: $10^3 \times 10^5 = ?$
Solution: $\log(10^3) + \log(10^5) = 3 + 5 = 8$
antilog(8) $= 10^8 = 100\,000\,000$.

That was like cracking a peanut with a sledge hammer, but it indicates how we operate with integer (whole number) powers of 10. We follow the same format even if we do not have integer powers of 10. We add the logs of the numbers to be multiplied and then find the antilog. But now we have to find the logs and antilogs from a table, calculator, or spreadsheet. No problem. Here you will see that scientific notation is really a helpful time-saver. Let's try it.

Find $\log(20)$.
First, write this in scientific notation:
$20 = 2.0 \times 10^1$
$\log(20) = \log(2.0 \times 10^1 = \log(2.0) + \log(10^1)$

The last step is by the rule of exponents that says the log of a product is the sum of the logs.

So, now we have $\log(20) = \log(2.0) + \log(10^1)$. Look at the upper right-hand corner of the 10 and you will see its log. The log of 10^1 is 1. So,

$$\log(20) = \log(2.0) + 1.$$

Using this technique, all we need to know are the logs between 1 and 10.

And now we introduce two new names. The $\log(2.0)$ is called the *mantissa*. You want to know why? . . . I'll tell you why. **Tradition!** There is a table of mantissas on the next page (Figs. 1-2 and 1-3) all worked out for you for numbers from 1.0 to 9.9. The power of 10 is called the *characteristic*. Same reason.

$$\log(20) = \text{mantissa} + \text{characteristic} = \log(2.0) + 1$$

Now I hope you can see the beauty and the mystique of writing numbers in scientific notation. To find the characteristic all you do is look at the upper right-hand corner of 10 to get the power of 10. To get the mantissa all you do is consult the Table or your calculator or computer. (Actually most calculators give you the result of adding the mantissa and the characteristic, and now you understand what they do.)

$$\log(20) = 0.3010 + 1$$

Sometimes it is more convenient to leave it like this instead of adding the numbers to get 1.3010. That way you can easily account for the powers of 10 because they are well removed from the mantissa. But, if you add the numbers, the numbers to the left of the decimal are the powers of 10 (the characteristic) and the ones to the right constitute the mantissa.

Use the worksheet LOGTABLE or a calculator for Self-Test 1-7.

Self-Test 1-7	
(a) log(10)	(f) $10^{-6}/10^3 =$
(b) log(1000)	(g) $(10^4)^3 =$
(c) 10 000 = 10^x ; x = ?	(h) $(10^8)^{1/2} =$
(d) 1 = 10^x ; x = ?	(j) 350: characteristic:_____ mantissa:_____
(e) $10^3 \times 10^5 =$	(k) 0.417: characteristic_____ mantissa_____

LOGTABLE Worksheet

Start your spreadsheet in the usual way. Open the file LOGTABLE and you should see something like the screens shown in Fig. 1-2 and 1-3.

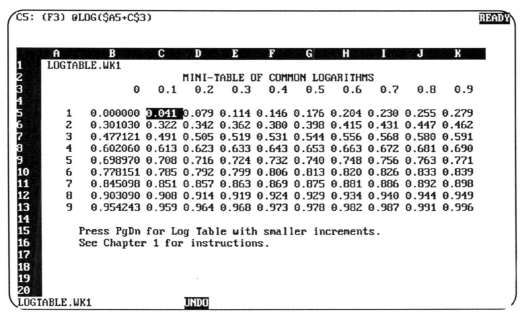

Fig. 1-2. Home screen of the LOGTABLE worksheet, in a Lotus *1-2-3* format.

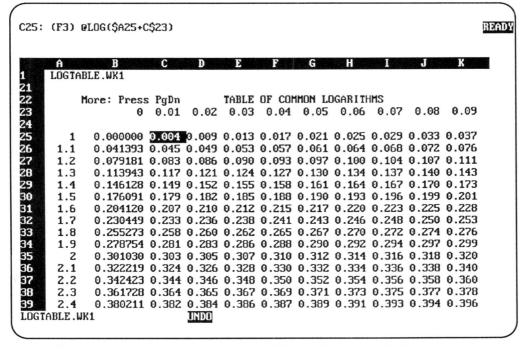

Fig. 1-3. Second screen of LOGTABLE. To see this press PgDn.

Wow! You deserve a break!

And now for the break I promised you. It's time for our ever-popular feature, *Culture Corner*, brought to you today by the Napier Corporation, producers of the world's finest logarithms. Don't just ask for common logs, ask for Napier's logs! They're all natural. While you're resting, consider the following Table:

Table 1-8. Space and Time		
Physical Parameter	*Distance or Time*	*Log of Ratio between Steps*
Smallest visible distance	0.1 mm	—
Human dimensions	1 m	4
Objects in landscape	10 km	4
Diameter of Earth	1.2×10^4 km	3
Earth to Sun	1.5×10^8 km	4
Sun to Sirius	10^{14} km	6
Size of galaxy	10^{18} km	4
Distance to nearest galaxy	10^{19} km	1
Size of universe	10^{23} km	4
Shortest time interval distinguishable by ear	0.1 sec	—
One day	10^5 sec	6
Human life (100 years)	10^9 sec	4
Human civilization	10^4 years	2
Development of humans, significant change of Earth's surface	10^6 years	2
Development of mammals, age of Rocky Mountains	10^8 years	2
Table continues on p. 13.		

Table 1-8. Continued.		
Physical Parameter	*Time*	Log of Ratio Between Steps
Age of many rocks	2×10^9 years	<1
Age of Earth	$\approx 4.5 \times 10^9$ years	<1
Age of universe	1 to 2×10^{10} years	<1

If the age of the universe is taken as one day, human beings have existed only for the last ten seconds.

LOG LOG Operations

Now that you've had your break, let's quickly finish up. Repeated log operations are useful once in a while. For example, one often encounters an exponential which has its exponent as a power, like the familiar Gaussian function,

$$y = \exp[-x^2/s].$$

In this expression s is a constant called the standard deviation, and often written as the Greek symbol σ. We will stick with the Latin s. Taking the natural log of both sides we obtain,

$$\ln(y) = -\frac{x^2}{s}$$

In this case we cannot take the natural log again because taking the logarithm of a negative number is a No-No (see General Warning #2, p. 5). But, if we have something like,

$$y = \exp[x^2/s]$$

then,

$$\ln(y) = \frac{x^2}{s}$$

and,

$$\ln[\ln(y)] = \ln(x^2/s)$$

so,

$$\ln[\ln(y)] = 2\ln(x) - \ln(s).$$

The second logarithm operation does not have to be to the same base as the first; one could have

$$\log[\ln(y)] = 2\log(x) - \log(s).$$

This may seem a bit convoluted but it is useful when one has something raised to a power which is itself raised to a power (which also may be raised to a power, *ad infinitum*).

Summary

Let's see what we've done so far. The logarithm of a product of two or more numbers is the sum of the logarithms:

$$\log(ABC) = \log(A) + \log(B) + \log(C).$$

The logarithm of a quotient of two numbers is the difference of their logarithms:

$$\log(A/B) = \log(A) - \log(B).$$

The logarithm of any power of a number is the power multiplied by the logarithm of the number. The power can be positive, negative, integer, or a fraction so this includes taking roots:

$$\log(A^n) = n\log(A)$$

It is important to write numbers in scientific notation to facilitate taking logs, to keep track of zeros, to compare very large and very small numbers easily, and to save printer's ink.

Let's look at some examples of calculations done in a neat form so you can see what is going on. Napier would have called these canonical forms if he would have written in English instead of Latin. *Omnia forma in tres partes divisa est.*

Multiplication
$543 \times 1\,210 = $ _____ ?
Write in scientific notation:
$(5.43 \times 10^2)(1.21 \times 10^3) = $ _____
Take logs:
$\log(5.43 \times 10^2) + \log(1.21 \times 10^3) = $ _____
Find and add logs:
$(0.7348 + 2) + (0.0828 + 3) = 0.8176 + 5$
Find antilog:
$\text{antilog}(0.8176 + 5) = 6.57 \times 10^5$

Division

0.0477/3930 = _____ ?
Write in scientific notation:
$(4.77 \times 10^{-2})/(3.93 \times 10^{3}) =$ _____
Take logs:
$\log(4.77 \times 10^{-2}) - \log(3.93 \times 10^{3}) =$ _____
Find and subtract logs:
$(0.6785 - 2) - (0.5944 + 3) = 0.0841 - 5$
Find antilog:
$\text{antilog}(0.0841 - 5) = 1.214 \times 10^{-5}$

Powers and Roots

$(653)^{1.2} =$ _____ ?
Write in scientific notation,
$(6.53 \times 10^{2})^{12} =$ _____
Take log
$1.2 \log(6.53 \times 10^{2}) =$ _____
Find log and multiply by power or root.
$1.2 (0.8149 + 2) = 0.9779 + 2.4$
You need an integer for the characteristic, so borrow 0.6 from the mantissa and add it to the characteristic to get,
$0.3779 + 3.$
$0.9779 + 2.4 = 0.3779 + 3$
Find antilog
$\text{antilog}(0.3779 + 3) = 2.39 \times 10^{3}$ (approximately)

EXAMPLE
The Even-Tempered Scale

As we mentioned before, the twelfth root of 2 is very important in music because it is the basis of the even-tempered musical scale. This chromatic scale has been preferred by composers since the nineteenth century. The classical problem was to find $2^{1/12}$, that is, a number which when multiplied by itself 12 times, equals 2. Now we can do this easily with logarithms.

$$\log(2^{1/12}) = (1/12)\log(2)$$
$$= (1/12)(0.30103)$$
$$= 0.025086, \quad \text{approximately}$$
$$2^{1/12} = \text{antilog}(0.025086) = 1.059463, \text{ approximately}$$

So What?

So, now we have the rudiments of a streamlined, easy way to do difficult calculations. But what does this have to do with the decibel?

Well, for one thing, millions of years ago the creatures who could hear a tiger creeping up on them during a thunderstorm were the ones who survived and multiplied. The creatures who couldn't hear this well were the ones who enabled the tigers to survive and multiply. Our ancestors bequeathed to us a hearing system with a logarithmic response so that we can hear the faintest sounds with great sensitivity without having our heads blown off by sounds that are only a million or a billion times more powerful. The logarithmic response of our hearing system means we have a large *dynamic range*. More about that term later.

Although people cannot sense absolute intensity or sound pressure, they can readily sense *changes* in intensity or sound pressure. If there are two sound intensities, say I_1 and I_2, people do not sense the difference of the intensities, $I_1 - I_2$. People sense, approximately, the *difference of the logs,* $\log(I_1) - \log(I_2)$. By the rules of logarithms this means that people sense the log of the ratio,

$$\log\left(\frac{I_1}{I_2}\right).$$

Patience. We're almost there. But first, let's get a firm idea of what we mean by *intensity*. As used here, intensity is a word that means power per unit area. (Power *density* is power per unit volume.)

Power is energy per unit time. So, intensity is energy per unit time per unit area. The common unit of power is the Watt, and area is measured in square meters. This definition is shown graphically in Fig. 1-4. It is important to distinguish between intensity (an objective physical quantity) and *loudness* (a subjective perception). We'll discuss this in Chapter 2.

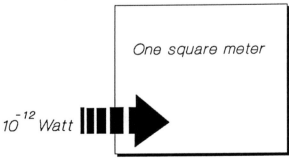

Fig. 1-4. Intensity is measured in terms of a specified power incident on a given area. The area can be an imaginary area in free space or an actual surface. The intensity shown here is 10^{-12} Watt/m^2, which is the threshold of normal human hearing. This intensity can also be expressed as 10^{-16} Watt/cm^2.

Powers of Two

Once upon a time, long ago and far away in India's sunny clime, there lived a wealthy maharajah who was more fond of chess than the average person. Indeed, the maharajah was so pleased with the game that he wished to reward the inventor. (I beg your forgiveness if the following translation from the original Sanskrit has lost some of its flavor.)

And so it came to pass before a fortnight had elapsed that the inventor was called before His Excellency. The inventor was young, handsome, intelligent, and modest, not unlike your humble storyteller, Dear Reader.

"I am inordinately enthralled with the game you have invented," said the maharajah, "and therefore I wish to bestow upon you honors and riches. How about half of my kingdom, the hands of my daughters in marriage, and so forth?"

"Aw shucks, your Lordship," replied the inventor, "all I want is some rice. Suppose you just give me one grain of rice for the first square of the chessboard, two for the second, four for the third, eight for the fourth, and so on up through the sixty-fourth square. That's all I want."

"So be it! You shall have all you have requested," said the maharajah, and he forthwith sent his commissary officer out for a bag of rice.

And truly it is written: There was not enough rice in the world for the sixty-fourth square. Table 1-9 shows how it goes.

Table 1-9. Grains of Rice on a Chessboard		
Square Number	Power of Two	Grains of Rice on Square
1	2^0	1
2	2^1	2
3	2^2	4
4	2^3	8
64	2^{63}	$9.223371987 \times 10^{18}$

The total number of rice grains for all 64 squares is $2^{64} - 1$, which is

$$18\ 446\ 744\ 073\ 709\ 551\ 615.$$

Another version of this tale, of doubtful authenticity, attributes the invention of chess to the Grand Vizier Sissa Ben Dahir of the court of the Indian king, Shirhâm. Sissa is said to have asked for grains of wheat instead of rice; this much wheat would cover the surface of Planet Earth to a depth of approximately 1.27 mm (0.05 inch).

Advice to Youth

The science of calculation also is indispensable as far as the extraction of the square and cube roots; algebra as far as the quadratic equations; and the use of logarithms are often of value in ordinary cases; but all beyond these is but a luxury; a delicious luxury indeed; but not to be indulged in by one who is to have a profession to follow for his substinence.

Thomas Jefferson

Answers to Self-Tests

Self-Test 1-1	
(a) 10^8	(d) 10^{18}
(b) 10^{15}	(e) 10^{15}
(c) 10^{11}	(f) 10^{10}

Self-Test 1-2	
(a) 10^5	(d) 101
(b) 10^3	(e) 102
(c) 10^{-1}	(f) 10^4

Self-Test 1-3	
(a) 1/10 000 = 0.0001	(d) 100 000 = 10^6
(b) 1/1 000 000 = 0.000001	(e) 100 = 10^2
(c) 1 /1 000 = 0.001	(f) 1 /100 = 0.01

Self-Test 1-4	
(a) 4	(d) 10^{12}
(b) −3	(e) 10^{-6}
(c) 10^8	(f) 10^{-10}

Self-Test 1-5	
(a) 10^3	(d) 10^{-1}
(b) 10^4	(e) 10^{-8}
(c) 10	(f) 10^{22}

Self-Test 1-6	
(a) 1.25×10^2	(d) 1.42×10^{-4}
(b) 4.55×10^{-1}	(e) 26 700
(c) 1.678×10^6	(f) 3 555

Self-Test 1-7	
(a) 1	(f) 10^{-9}
(b) 3	(g) 10^7
(c) 4	(h) 10^4
(d) 0	(i) characteristic: 2 mantissa: 0.5441
(e) 10^8	(j) characteristic: −1 mantissa: 0.6201

Digging Deeper

Powers of Ten, Philip Morrison, Phylis Morrison, and The Office of Charles and Ray Eames (W. H. Freeman and Company, San Francisco, 1982). ISBN 0-7167-1409-4. This beautiful book, based on the film of the same name, zooms photographically from the edges of the universe down to the size of quarks.

Powers of Ten, a 9.5-minute film, was made for IBM by The Office of Charles and Ray Eames. A 21-minute version featuring Gregory Peck is available on VHS video tape (*Powers of Ten: The Video,* catalog number VT 110) from The Astronomical Society of the Pacific, 390 Ashton Avenue, San Francisco, CA 94112. Phone (800) 335-2624; FAX (415) 337-5205; e-mail asp@stars.sfsu.edu. The short version is available as catalog number CM274 from Insight Media, 2162 Broadway, New York, NY 10024. Phone (212) 721-6316; FAX (212) 799-5309.

Chapter 2

ABCs of dBs

The Bel

We said that humans hear approximately like $\log(I_o/I_r) = \log(I_o) - \log(I_r)$. Intensity is measured in Watts/m^2 for all kinds of waves, including sound and light. Intensity is an objective physical quantity measured in physical units; loudness is the physiological and psychological correlate of intensity and is subjective. Let's examine this in more detail in terms of the stimulus (input) to a system and the system's response to the stimulus (output). If the output really depends on the input then we can find some functional relationship between them.

For example, suppose we find experimentally that the output of a system is 5 times the input. Let y be the output and x be the input, so we can write $y = 5x$. In words, the output is always five times the input; the output is directly proportional to the input. This is shown graphically in Fig. 2-1.

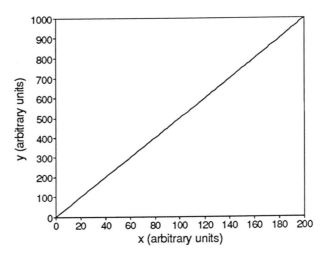

Fig. 2-1. Output is proportional to the input. The graph is linear in x and y.

We call this a linear function because it depends only on the first power of x and its graph is a straight line. Another way of looking at this is to say that the ratio of the output to the input is a constant, $y/x = 5$. If x is 10 then y is 50, if x is 15 then y is 75, and so on.

This graph quickly becomes very large. If the input has a large dynamic range then the scale of the graph must be modified so that we can see all of the values, but we cannot see the small values very well in the presence of large values. If the input becomes very large the system response must become very large; there is a limit to everything (except taxes) and the system will become overloaded at large inputs, with usually dire consequences. Now suppose that the y-axis has a logarithmic scale, so that $y = \log(5x)$. This scale is shown on the right-hand axis in Fig. 2-2.

Notice how the logarithmic scale increases as the input increases. The increase in y is progressively less as the input increases. In such a system you can measure small inputs and large inputs without overloading the system. This is the beauty and mystique of the logarithmic hearing provided to us through the courtesy of our ancestors who were given it or had sense enough to develop it before the tigers ate them.

We define the unit Bel as the log of the ratio of a sound intensity (I_o) to a reference sound intensity (I_r). The commonly-used references are in Table 2-3 on page 28.

$$\text{Bel} = \log\left(\frac{I_o}{I_r}\right)$$

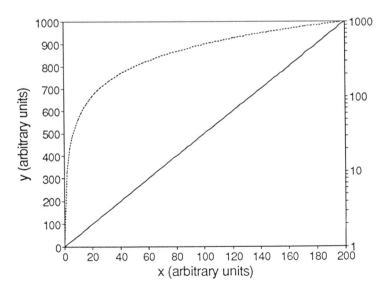

Fig. 2-2. *Left axis:* Output is a linear function of the input (straight line). *Right axis:* Output is a linear function of the input (curve) but is scaled logarithmically.

Bels are positive when $I_o > I_r$, Bels are zero when $I_o = I_r$, and Bels are negative when $I_o < I_r$. This is all in accordance with the rules of exponents. The Bel is named in honor of the inventor of the telephone, Alexander Graham Bell (1847-1922). He also put bells on telephones. Bels aren't used much but decibels are used over and over. They never seem to wear out.

The Decibel

The prefix *deci-* is the metric system scale factor that represents $1/10 = 0.1 = 10^{-1}$ so that a decibel is one-tenth of a Bel. Normal humans can distinguish a change of about a one-tenth of a Bel, or a decibel. The decibel is abbreviated as dB.

$$dB = 10 \log_{10} \frac{I_o}{I_r}$$

Before we proceed further with the decibel we must digress for a few moments to clarify how we measure intensity, pressure, power, and voltage. All of these quantities are generally changing in time; they may even be totally unpredictable, like random noise. No problem. The key to comparing complicated time-varying quantities such as these is through the concept of rms (pronounced "arr em ess").

RMS

The abbreviation rms means root-mean-square, or the square root of the mean square. The concept of rms enables us to compare decibels in different types of waves such as speech, music, noise, pure tones, and pulses. They say you can't compare apples and oranges, but we *can* compare apples and oranges. The unifying common denominator is rms. In order not to interrupt our stream of consciousness and get on with decibels we have sequestered the details of rms in Appendix 2 where they are safe from prying eyes. Only you, Dear Reader, are not forbidden to examine them.

From now on, all intensities will be time-averaged intensities and all pressures will be rms pressures, where the average is taken over a long time interval. Similarly, voltages will be rms voltages unless otherwise noted. Almost all common voltmeters are calibrated to read rms values for a sine wave and will give erroneous readings if other waveforms are used. A so-called *true rms voltmeter* will measure, by thermal means or digital computation, the rms value of any waveform including a completely random wave such as noise. Digital instruments can be programmed to compute the rms value. Instruments intended for acoustic measurement almost always use rms values. The rms value is not the only useful parameter for characterizing complicated waves; others are also discussed in Appendix 2.

Sound level meters provide a "quasi-rms" value of a time-varying sound pressure. The time constants used in sound level meters operating in accordance with ANSI S1.4-1983 (American National Standard Specification for Sound Level Meters) are shown in Table 2-1.

Table 2-1. Sound Level Meter Time Constants (ANSI S1.4-1983)	
Label	r m s Detector Time Constant
Slow (S)	1 s
Fast (F)	0.1 s
Impulse (I)	0.035 s = 35 ms

Some sound level meters have a peak-hold feature which captures the instantaneous greatest sound pressure level. The peak reading should not be confused with the maximum reading, which refers to the detected level, weighted A, B, or C, and Slow, Fast, or Impulse. Data acquired by sound level meter readings should include the complete description as, for example, 60 dB A-weighted, Fast.

Exponents Revisited

Because decibels are logarithmic and logarithms are simply exponents, we will briefly review (once again, but with more feeling!) the rules of exponents. If you had hit the fast-forward button back there in Chapter 1 it won't do you any good now. (You know who you are.) When powers of a given quantity are multiplied the following rule applies:

$$x^m x^n = x^{m+n}.$$

For example,

$$x^3 x^4 = x^7.$$

When dividing powers of a given quantity,

$$\frac{x^m}{x^n} = x^{m-n}.$$

For example,

$$\frac{x^5}{x^3} = x^2.$$

An exponent that is a fraction corresponds to a root.

$$x^{1/2} = x^{0.5} = \sqrt{x}.$$

Any quantity x^m that is raised to the power n can be expressed as,

$$(x^m)^n = x^{mn}.$$

For example,

$$(x^2)^3 = x^2 x^2 x^2 = x^6.$$

A special case should be noted:

$$x^0 = 1.$$

In fact, any number raised to the zero power is equal to one. Does any of this sound familiar?

Logarithms Revisited

Logarithms are simply exponents. The base of a logarithm is just the number that is raised to the power. For example, consider $10^3 = 1000$. The base is 10 and 3 is the logarithm of 1000 to the base 10. For decibels we use base 10 but other bases are also important so we will briefly discuss logarithms in more generality than in Chapter 1. Consider the expression,

$$x = b^y.$$

The number b is the base and y is the exponent; we say that y is the logarithm of x to the base b:

$$\log_b(x) = y \quad \text{because} \quad x = b^y.$$

Conversely, the antilogarithm of y to the base b is x,

$$\text{antilog}(y) = x.$$

The base 10 is the base of common logarithms and the base $e = 2.71828\ldots$ is the base of natural logarithms. Therefore, for common logs,

$$y = \log_{10}(x) \quad \text{means} \quad x = 10^y.$$

Very often we simply write $y = \log(x)$ for common logarithms. From now on when we write log we will mean $\log_{10}$ and when we write ln we will mean natural logarithms. For natural logarithms,

$$y = \ln(x) \quad \text{means} \quad x = e^y.$$

Examples

$\log(52) = 1.716$

$\text{antilog}(1.716) = 10^{1.176} = 52.$

$\ln(52) = 3.951$

$\text{antiln}(3.951) = e^{3.951} = 52.$

Sometimes it is important to convert from common logs to natural logs, and vice versa. This can be done as follows:

$\ln(x) = 2.302585 \log(x).$

$\log(x) = 0.434294 \ln(x)$

Suppose that you know $\log(53.1) = 1.7251$. What is $\ln(53.1)$?

$\ln(53.1) = 2.3026 \log(53.1)$

$\ln(53.1) = 2.3026 * 1.7251 = 3.9722$

Conversely, if you know $\ln(53.1) = 3.9722$, then

$\log(53.1) = 0.434294 \ln(53.1)$

$\log(53.1) = 0.434294 \times 3.9722 = 1.7251$

Some useful properties of logarithms are listed here for your convenience:

$\log(ab) = \log(a) + \log(b)$

$\log(a/b) = \log(a) - \log(b)$

$\log(a^n) = n \log(a)$

$\log(10) = 1$

$\ln(e) = 1$

$\ln(e^n) = n$

$\ln(e^x) = x$

$\log(x) = \ln(x)/\ln(10)$

$\exp[\ln(x)] = x$

$\ln(1/x) = \ln(x^{-1}) = -\ln(x)$

$10^{\log(x)} = x$

Sound Intensity Level, dB(SIL)

Sound intensity is measured in power (Watts) per square meter, W/m^2. In other words, intensity is just a certain amount of power per unit area. A small amount of power concentrated in a small area corresponds to a high intensity. If the same amount of power is spread out over a large area the intensity is correspondingly smaller. Think of a magnifying glass concentrating the power from the sun into a small spot; the magnifying glass has certainly increased the light intensity but has not increased the power.

The human ear can detect an astonishingly wide range of sound intensities. The faintest sounds the normal human ear can detect at a frequency of 1000 Hz correspond to an intensity of about 10^{-12} W/m^2. This is commonly taken to be the threshold of hearing. At the other end of the scale the loudest sounds that the ear can withstand before pain begins is about 1 W/m^2; this is called the threshold of pain. The range of intensity is therefore about 10^{12} or a thousand billion. We define the intensity level by,

$$dB(SIL) = 10 \log(I_o/I_r).$$

I_r is the reference intensity which is commonly taken to be the threshold of hearing, 10^{-12} W/m^2, and I_o is intensity in W/m^2 at the level in dB(SIL). All of these intensities are rms values. Sometimes you will see dB(SIL) written as dB(IL). By the rules of logarithms the log of a ratio can be written as the difference of two logs,

$$dB(SIL) = 10 \left[\log(I_o) - \log(I_r)\right].$$

In dB the reference intensity is given by,

$$dB(SIL) = 10 \log(I_r/I_r) = 10 \log(1) = 0 \text{ dB(SIL)}.$$

At the upper limit of hearing the threshold of pain is,

$$dB(SIL) = 10 \log(1/10^{-12}) = 10 \log(10^{12}) = 120 \text{ dB(SIL)}.$$

Table 2-2 gives some typical intensity levels.

Table 2-2. Sound Intensity Levels for Common Sources, dB(SIL)	
Nearby jet plane	150
Machine gun; jackhammer	130
Rock concert; siren	120
Power mower; subway	100
Heavy traffic	80
Vacuum cleaner	70
Normal conversation	50
Mosquito buzzing	40
Whisper	30
Rustling leaves	10
Threshold of hearing	0

Negative decibels simply mean that the sound intensity or pressure is less than the corresponding reference.

For example, how many dB(SIL) correspond to a sound intensity of 2×10^{-14} W/m^2 ?

$$dB(SIL) = 10 \log(2 \times 10^{-14}/10^{-12})$$

$$= 10 \log(2 \times 10^{-2}) = -17 \, dB$$

Prolonged exposure to high intensity levels may produce damage to the human ear. Such exposure is known to cause deafness in hamsters, too. Experimental evidence indicates that high noise levels may contribute to high blood pressure, anxiety, and nervousness. Protection such as ear plugs are suggested for intensities greater than 90 dB.

For a change, suppose we know dB(SIL) and we need to know the corresponding sound intensity. This is important in many situations, but particularly when we need to add or subtract decibels because we *can't* add or subtract decibels directly. We must add or subtract intensities.

EXAMPLE:
What intensity I_o corresponds to 80 db(SIL)?

$dB(SIL) = 80;$ $I_o = ?$

$80 = 10 \log(I_o/I_r)$

$8 = \log(I_o/10^{-12})$

Now take the antilog of both sides (raise both sides to the base 10, remember?)

antilog$(8) = 10^8 = I_o/10^{-12}$

and finally multiply both sides by 10^{-12} to find I_o,

$I_o = 10^8 \times 10^{-12} = 10^{-4} \, W/m^2$

So, we have found that 80 dB(SIL) corresponds to 10^{-4} W/m^2. And now for some dB practice. The answers to Self-Test 2-1 are at the end of this chapter.

Self-Test 2-1	
(a) I $= 2.5 \times 10^{-11}$ W/m^2 dB(SIL) =	(c) dB(SIL) $= 88$ I = _____ W/m^2
(b) I $= 7.9 \times 10^{-14}$ W/m^2 dB(SIL) =	(d) dB(SIL) $= -16$ I = _____ W/m^2

Sound Pressure Level, dB(SPL)

We define the decibel in terms of sound pressure level as,

$$dB(SPL) = 20 \log (p_o/p_r) .$$

The factor 20 appears because intensity is proportional to the square of the pressure. (All pressures are rms values.) To clarify this, since I is proportional to p^2, $I = kp^2$, then $I_o/I_r = k(p_o)^2/[k(p_r)^2] = (p_o/p_r)^2$ so

$$dB(SPL) = 10 \log (p_o/p_r)^2$$

By the rule of logarithms, $n \log(x) = \log(x^n)$, so this becomes

$$dB(SPL) = 20 \log (p_o/p_r)$$

Table 2-3 summarizes the commonly accepted reference intensity and pressure. (Recall 1 N/m^2 =1 Pascal)

Table 2-3. Zero dB: Standard Reference Intensity and Pressure ·	
Sound Intensity Level	Sound Pressure Level
10^{-16} W/cm^2	2×10^{-4} dynes/cm^2
10^{-12} W/m^2	2×10^{-5} Pascals

Let's try some more examples of dB(SIL) and dB(SPL) with the standard references. The answers to Self-Test 2-2 are at the end of this chapter.

Self-Test 2-2	
(a) $I_o = 9.1 \times 10^{-10}$ W/m^2 dB(SIL) =	(c) dB(SIL) = 95 I_o = _____ W/m^2
(b) $p_o = 4.0 \times 10^{-4}$ P $\qquad$ a dB(SPL) =	(d) dB(SPL) = 66 p_o = _____ Pa

Loudness and Loudness Level

The *phon* is the unit of loudness level. It is defined for pure tones (sine waves) as the intensity in dB of a 1 kHz tone that sounds equal in loudness to the given tone. Curves of constant loudness level have been measured for normal hearing many times. For examples, see Digging Deeper at the end of this chapter.

The curve of constant loudness level that touches the 40 dB line at 1000 Hz is called the 40 phon contour and is the dB(A) weighting curve, the contour at 70 phons is the dB(B) weighting curve, and the contour at 100 phons is the dB(C) weighting curve. Standard weighting for the dB(A), dB(B), and dB(C) scales is given in American National Standard for Sound Level Meters, ANSI S1.4-1971 (R1976).

The *sone*, a subjective unit of loudness, is defined as the loudness of a 1 kHz pure tone at an intensity level of 40 dB. The experimental relation between sones and phons is highly nonlinear (Fig. 2-3).

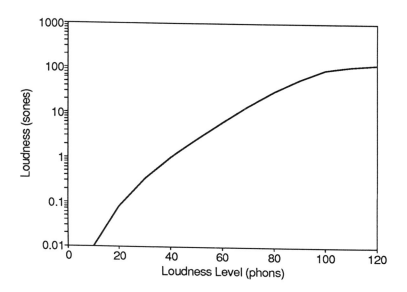

Fig. 2-3. Experimental relationship between loudness and loudness level. At 1 kHz, 20 phons = 20 dB, 40 phons = 40 dB, 60 phons = 60 dB.

Save the Forests: Use Decibel Graphs

The normal human hearing response has a range from −10 to +130 dB(SIL), or a range of 140 dB(SIL). Using the dB scale the vertical axis is a convenient size to fit on an ordinary piece of paper. However, suppose that instead of plotting the vertical axis in dBs we plot it linearly at a scale of, say, one inch =10 times the reference. In other words, on the linear scale two inches =20 times the reference, three inches = 30 times the reference, and so on.

For ten times the reference, dB = 10 log($10I_r/I_r$), or dB = 10 log(10) = 10 dB (SPL). Similarly, 20 times the reference corresponds to 13 dB, 100 times the reference corresponds to 20 dB, and on up to 10^{14} times the reference which corresponds to 140 dB. This is summarized in Table 2-4. What is the total size of the vertical axis?

$$\frac{10^{13} inches}{(12\ inches/foot)(5280\ feet/mile)} = 1.578 \times 10^9\ miles$$

There is not enough paper in the world for this graph; it would take 2.4 hours for light to go that distance. The distance from the earth to the sun is only about 92 million miles, or 8.2 minutes for light. Decibels certainly save paper and paper is made from wood. Therefore decibels save the forests!

Table 2-4. 1 inch = 2.54 cm = 10 Times Reference	
dB(SIL)	*Inches*
10	10^0 inches=10^1 I_r
20	10^1 inches=10^2 I_r
30	10^2 inches=10^3 I_r
40	10^3 inches=10^4 I_r
50	10^4 inches=10^5 I_r
60	10^5 inches=10^6 I_r
140	10^{13} inches=10^{14} I_r

Designer dB Graphs vs. Generic dB Graphs

When making measurements as a function of frequency it is common practice to take data at evenly spaced points, say every 10 Hz or 100 Hz. On a logarithmic frequency scale this results in data being spread out at low frequencies and squeezed together at high frequencies. Here are some tips on plotting data that are evenly spaced on a logarithmic scale, for cases when data are measured over full decades and when data are measured over parts of decades.

Logarithmic Graphs, Evenly Spaced Points Between Decades

Suppose that you want to pick n evenly spaced points along the horizontal axis of semi-log graph paper. Each value will be larger than the previous one by a factor of the nth root of 10, i.e., the points are given by,

$$10^{1/(n-1)}$$

In the exponent n is the desired number of points per decade. For example if you are plotting frequencies between 10 kHz and 100 kHz and you want six points, you calculate $10^{1/5}$ (that is, the 5th root of 10, which is 1.5849) and the frequency spacing will then be:

10, 15.85, 25.12, 39.81, 63.09, and 100 kHz.

Logarithmic Graphs, Evenly Spaced Points Not Involving Decades

The method described above works with decades. Now suppose that you want evenly spaced intervals not involving decades; in this case the basic multiplier is given by,

$$(\text{upper limit/lower limit})^{1/(n-1)}$$

For example, for ten points between 50 and 10,000 Hz, the multiplier is

$$(10,000/50)^{1/(10-1)} = 1.8016.$$

Therefore the frequency spacing is as follows:

50, 90.08, 162.3, 292.4, 526.81, 949.12, 1709.98, 3080.78, 5550.47, 10,000 Hz.

OSHA Standards

Table 2-5 gives the Occupational Safety and Health Administration standards for the maximum allowable exposure to acoustic energy.

Table 2-5. Damage Risk Table	
dB(A) Scale*	Hours/Day
90	8
92	6
95	4
97	3
100	2
102	1.5
105	1
110	0.5
115	0.25 (or less)

*Note: The dB(A) scale is referred to the 40-phon equal loudness contour. The sound level meter is set to Slow (S) response, which corresponds to a detector time constant of one second.

When the daily noise exposure is composed of two or more periods of noise exposure of different levels, their combined effect should be considered, rather than the individual effect of each.

For example, let t_i be the total time of exposure at noise level i and let T_i be the total time of exposure permitted at that level. If the sum of the fractions,

$$\frac{t_1}{T_1} + \frac{t_2}{T_2} + \frac{t_3}{T_3} + \cdots$$

exceeds unity then the mixed exposure should be considered to exceed the OSHA limit value.

Decibel Addition: Shoot-Out at the dB Corral

It is important to be able to add and subtract sound levels, voltages, and power expressed as decibels. Since they are exponents they do not add or subtract like ordinary numbers.

Two outlaws meet at the old dB Corral in Tombstone, Arizona. Each outlaw has identical .45 caliber Colt revolvers. Everybody knows that each pistol, fired alone, is equivalent to 80 dB(SPL) at a distance of ten paces. When both pistols are fired together, how many dB are measured at that distance? Should the outlaws be arrested for violating OSHA Standards?

An angry crowd has gathered; they are upset because they know that violation of OSHA Standards could give Tombstone a bad name. Some of the citizens of Tombstone yell "Hang them!" because they mistakenly think that 80 dB + 80 dB = 160 dB. Fortunately, Wild Bill Gain and Calamity Jane Decibel, the most famous audiologists in the Old West, come out of the Silver Dollar Saloon and they calm the angry citizens with the following explanation:

$$80\,dB\ +\ 80\,dB\ =\ ?$$

We cannot add decibels directly. We must add the sound pressures from the two guns. First we must find what sound pressure p_o corresponds to 80 dB for one gun, $80 = 20 \log(p_o/p_r)$, where the reference sound pressure is $p_r = 2 \times 10^{-5}$ Pa.

$$80/20\ =\ 4\ =\ \log(p_o/p_r)$$

Next take the antilog of this equation, recalling that $10^{\log(x)} = x$, and obtain,

$$10^4\ =\ p_o/p_r$$

$$p_o\ =\ 10^4 p_r$$

Now, since there are two identical revolvers, the total sound pressure is $2p_o$, or $2 \times 10^4 p_r$. We can use this to find the dBs that correspond to the two guns fired simultaneously.

$$dB\ =\ 20 \log(2 \times 10^4 p_r/p_r)$$

$$dB\ =\ 20 \log(2 \times 10^4)$$

$$dB\ =\ 2\,0 \log(2)\ +\ 20 \log(10^4)$$

$$dB\ =\ 6.02\ +\ 80$$

$$80\,dB\ +\ 80\,dB\ =\ 86.02\,dB$$

The crowd accepts Bill and Jane's explanation, but just to be sure they hang the outlaws anyway.

There are two morals to this story:

- Hanging outlaws always reduces noise level.
- Doubling the sound pressure always results in adding 3 dB(SIL) or 6 dB(SPL).

In other words, all you have to remember is that 2 plus 2 is 5 or 8. Similarly, reducing sound pressure by one-half always results in subtracting 3 dB(SIL) or 6 dB(SPL). In other words, half of 10 is 7 or 4. All of your life you have heard in dBs; now it's time to think in dBs. Table 2-6 should help you get an intuitive feel for logarithmic changes:

Table 2-6. Examples of Logarithmic Changes			
Power	dB Change	Pressure, Pa	dB Change
Original: 2 mW	0	1	0
Doubled: 4 mW	+3	2	+6
4 × : 8 mW	+6	4	+12
8 × : 16 mW	+9	8	+18
10 × : 20 mW	+10	10	+20
100 × : 200 mW	+20	100	+40
1000 × : 2000 mW	+30	1000	+60

More on Decibel Addition

We have seen that doubling sound pressure adds 6 dB(SPL) and doubling sound intensity adds 3 dB(SPL). Now let's turn our attention to the general problem of addition. Remember, we must add the sound pressures or intensities, not the decibels.

The first step, then, is to convert decibels to sound pressures or intensities. The second step is to add the pressures or intensities, and the third step is to convert the sum back into decibels.

Example:

Suppose we want to add 50 dB(SPL) and 60 dB(SPL). You will see that it doesn't make any difference what the reference is; the reference always cancels out so you do not have to put in its numerical value. Just carry the reference value along as a symbol like p_r .

$$50 \text{ dB(SPL)} = 20 \log(p_1/p_r) \qquad 60 \text{ dB(SPL)} = 20 \log(p_2/p_r)$$

$$50/20 = \log(p_1/p_r) \qquad 60/20 = \log(p_2/p_r)$$

Antilog both sides:

$$\text{antilog}(2.5) = (p_1/p_r) \qquad \text{antilog}(3) = (p_2/p_r)$$

$$316.23\, p_r = p_1 \qquad\qquad 1000\, p_4 = p_2$$

Add sound pressures:

$$p_1 + p_2 = 316.23\, p_r + 1000\, p_r$$

$$= 1316.23\, p_r$$

Convert back to dB(SPL)

$$50\ \text{dB(SPL)} + 60\ \text{dB(SPL)} = 20\ \log(1316.23\, p_r/p_r)$$

$$\text{dB} = 62.39\ \text{dB(SPL)}$$

ADD-DB, Worksheet for Decibel Addition

Addition of decibels is a repetitive task that can be accomplished conveniently with a simple computer program or with a spreadsheet. This is a great advantage because you can play with various values of decibels and quickly gain an intuitive feeling for how things go. After you start your spreadsheet retrieve the worksheet named ADD-DB. Your monitor screen should look something like Fig. 2-4 .

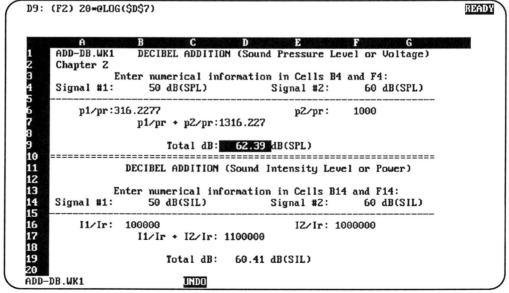

Fig. 2-4. Worksheet for decibel addition. The cursor is on Cell D9, and it displays the formula in this Cell at the upper left-hand corner of the screen.

In Fig. 2-4 Rows 4 through 9 perform addition for decibels in Sound Pressure Level or Voltage. As you saw in the numerical example the actual value of the reference does not matter except that both signals must have the same reference. The reference always cancels out.

The worksheet operates exactly the same as the calculations in the numerical example, except that as you enter the value of the decibels the results are automatically calculated. Move the cursor to Cell B6, F6, and D9 and observe their formulas.

Enter whatever values you want in Cells B6 and F6. Their sum appears instantly in Cell D9. In particular, try entering two identical values. What do you see? Of course, your see that the sum is 6.02 dB greater than either one. For example, $2 + 2 = 8.02$.

The decibel addition is also shown in the graph named *dB SPL or Volt*. You can access this graph in the usual way. All graphs are "live" which means that they are connected to their worksheets and instantly show changes in data. This graph should look something like Fig. 2-5.

Rows 12 through 20 provide decibel addition for Sound Intensity Level or Power. Simply enter the decibel values in Cells B15 and F15. The result is shown in Cell D20 and the graphical result is shown in the graph named *dB SIL or Power*.

Decibel Subtraction: Bill and Jane in Las Vegas

Bill Gain and Jane Decibel are in their private jet on their way from Tahiti to Monaco, having just completed a video tape of another program for "Lifestyles of the Rich and Famous Audiologists."

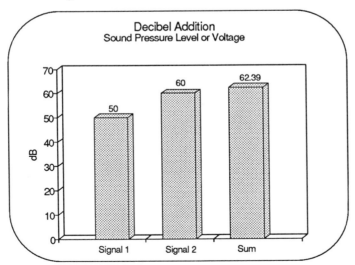

Fig. 2-5. The appearance of this graph depends on which spreadsheet you use. This format was produced in *Quattro Pro*.

Suddenly they receive a radio message calling them to an emergency assignment in Las Vegas. OSHA has threatened to shut down Nero's Palace because the noise of the roulette wheels at the gambling tables exceeds OSHA Standards in the Damage Risk Table.

Soon after landing in Las Vegas, Jane measures the noise level as 107 dB(SPL) when all roulette wheels are in operation. Bill measures a single roulette wheel as 65 dB(SPL) when all the others are stopped. The operators might have hidden some of their wheels in a back room; without counting them, how many roulette wheels does Nero's Palace have? How many did Jane and Bill shut down to comply with OSHA Standards? The answers to these and other exciting questions will now be revealed.

Approximate Method

107 dB(SPL) − 65 dB(SPL) =42 dB(SPL). Now, since 6 dB(SPL) corresponds to doubling sound pressure, 42/6 =7 doublings. If sound pressure is doubled 7 times, then $2^7 = 128$. This may be seen more clearly in Table 2-7.

Table 2-7. Roulette Wheel Noise Levels	
Number of Roulette Wheels	*dB(SPL)*
1	65
2	71
4	77
8	83
16	89
32	95
64	101
128	107

Referring to the OSHA Damage Risk Table on page 31, we see that 90 dB(SPL) is the maximum allowable for an 8-hour/day exposure. Therefore the maximum allowable number of roulette wheels is slightly more than 16, which corresponds to 89 dB(SPL). To be one dB on the safe side the number of roulette wheels to be shut down should be about $128 − 16 = 112$.

Accurate Method

Actually, doubling sound pressure is not exactly 6 dB(SPL). Recall log(2) = 0.3010, so 20 log(2) = 6.02 dB(SPL). That extra 0.02 dB(SPL) may not seem like much but it all adds up. Let's look at sound pressures.

$$65 \text{ dB(SPL)} = 2\ 0 \log(p_o/p_r)$$

Divide both sides by 20,

$$65/20 = 3.25 = \log(p_o/p_r)$$

Now antilog both sides of the above equation,

$$10^{3.25} = 1778.28 = (p_o/p_r)$$

So 65 dB(SPL) coresponds to $1,778.28 p_r$, which corresponds to one roulette wheel. In a similar way we find that 107 dB(SPL) corresponds to $10^{5.35} = 223,872.11 p_r$. Dividing $10^{5.35}$ by $10^{3.25}$ gives 102.1 (remember, dividing we subtract exponents). The result is $10^{2.1} = 125.89$ or 126 roulette wheels were in operation.

From the OSHA Table, we can only allow 90 dB(SPL) for 8 hours of exposure per day. 90 dB(SPL) corresponds to $10^{4.5} p_r$. Dividing $10^{4.5}$ by 103.25 gives $10^{1.25} = 17.78$ or 18 roulette wheels could be in operation at any given time.

So, $126 - 18 = 108$ roulette wheels had to be shut down. This made the operators of Nero's Palace a little happier than the approximate method. But, to be on the safe side again, after giving their recommendations Jane and Bill boarded their private jet and got out of town as soon as possible.

More on Decibel Subtraction

In general, if we want to subtract decibels we must subtract the sound pressures or sound intensities. As in the case of decibel addition, the reference will always cancel. First, convert decibels to their corresponding sound pressures. Second, subtract the pressures. Third, convert the resulting sound pressure back to decibels.

Example:
Suppose we want to subtract 74 dB(SIL) from 81 dB(SIL).

$$74 \text{ dB(SIL)} = 10 \log(I_1/I_r)$$
$$81 \text{ dB(SIL)} = 10 \log(I_2/I_r)$$
$$74/10 = \log(I_1/I_r)$$
$$81/10 = \log(I_2/I_r)$$

Antilog both sides:
$$\text{antilog}(7.4) = (I_1/I_r)$$
$$\text{antilog}(8.1) = (I_2/I_r)$$
$$25,118,864.3 \, I_r = I_1$$
$$125,892,541.2 \, I_r = I_2$$

Substract intensities:
$$I_2 - I_1 = 125,892,541.2 \, I_r - 25,118,864.3 \, I_r$$
$$= 100,773,676.9 \, I_r$$

Convert back to dB(SIL)

$$dB(SIL) = 10 \log(100{,}773.676.9 \, I_r/I_r)$$

$$81 \, dB(SIL) - 74 \, dB(SIL) = 80.03 \, dB(SIL)$$

Notice that the reference intensity always cancels itself.

SUB-DB, Worksheet for Decibel Subtraction

The worksheet SUB-DB is provided to automate the repetitive task of subtracting decibels, and to give you some practice so you can gain an intuitive feeling for the process. Open this worksheet and you should see something like Fig. 2-6.

Rows 5 through 10 perform subtraction of decibels for Sound Pressure Level or Voltage. As you have seen the value of the reference always cancels out in decibel subtraction. The important point is that both signals must have the same reference.

Enter some new values in cells B5 and F5. Their difference appears in cell D10. Remember that the larger value of the input signal must be placed in cell B5 and the smaller in cell F5. If you do not follow this instruction the spreadsheet will try to calculate the logarithm of a negative number (this is technically known as a No-No) and the message "ERR" will appear in cell D10. Move the cursor around to the other cells and you can see that this worksheet performs decibel subtraction exactly the same as the numerical examples.

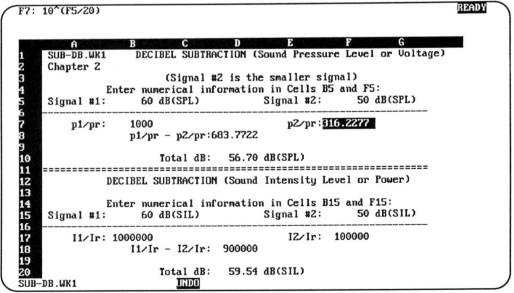

Fig. 2-6. Worksheet screen for decibel subtraction. The cursor is on cell F7, and it displays the formula in this cell at the upper left-hand corner of the screen.

The graph connected to Rows 5 through 10 is named *dB SPL or Volt* and you can see it on your monitor screen using the command appropriate to your spreadsheet. You should see something like Fig. 2-7.

Rows 15 through 20 in this worksheet perform decibel subtraction for Sound Intensity Level or Power. Again, move the cursor around this worksheet so you can see how the values are calculated. The graph connected with this part of the worksheet is named *dB SIL or Power*.

Reference Threshold Levels: dB(HL) and dB(SL)

Table 2-8 lists the frequency and dB(SPL) combinations considered to be the threshold for normal hearing according to the 1951 ASA Standard and the 1964 ISO Standard. Entries in the Table give 0 dB(HL) at the specified frequency. The term dB(HL) means Hearing Level for a normal subject.

The ISO Standard is in wider use. The audiometer generates different sound pressure levels at each frequency in the Table, thus compensating for the normal variation in sensitivity at different frequencies. The reference changes at each frequency. Increasing the dB(HL) output by, say 20 dB(HL), increases the pressure output by 20 dB(HL) at each frequency.

In using an audiometer it is very important to know to what Standard the audiometer was calibrated. Audiograms should be permanently labeled "1951 ASA" or "1964 ISO".

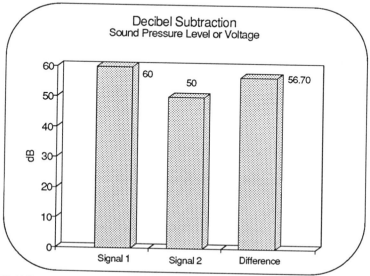

Fig. 2-7. When the SUB-DB worksheet is imported into *Quattro Pro* you can make a 3-D graph that changes instantly when the input signals change.

Table 2-8. Reference Threshold Levels		
Frequency, Hz	1951 ASA, dB(SPL)	1964 ISO, dB(SPL)
125	54.5	45.5
250	39.6	24.5
500	24.8	11.0
1000	16.7	6.5
2000	17.0	8.5
4000	15.1	9.0
8000	20.9	9.5

The term dB(HL) refers to the Standards in the Table for *normal* hearing. This means that a normal subject would just perceive the sound at the various frequencies. The term dB(SL) means Sensation Level, and refers to the *individual* subject, not the "normal" subject; dB(SL) means that the reference is the subject's own hearing, not the "normal" hearing reference.

Exponential Growth and Decay

Exponential growth and decay occur whenever the rate of change of a system parameter is proportional to the present state of the system. Because sound waves and electromagnetic waves are disturbances in space and time it is important to distinguish between spatial and temporal growth and decay.

Temporal growth and decay refer to the way something behaves in time, at one point in space. Temporal growth and decay are related to absorption which is related to reverberation time, as you will see later in Chapter 4. Spatial growth and decay refer to the way something behaves in space, at one point in time. Spatial growth and decay are related to absorption and amplification in traveling waves in things like lasers, fiber optic amplifiers, and microwave traveling wave tubes.

Temporal Growth: Parable of the Unicorns

To get a feeling for these processes, consider a herd of unicorns with unlimited food and water and no predators. If there are plenty of healthy boy and girl unicorns then the rate of increase in the number of unicorns will be proportional to the number of unicorns at any time. The more unicorns there are, the more there will be:

$$\frac{\Delta N}{\Delta t} = kN$$

In this equation ΔN is the increase during the time interval Δt, N is the number at a given time, and k is the constant of proportionality. Another way of looking at the herd is to re-arrange the above expression to say that the fractional increase is proportional to the time interval,

$$\frac{\Delta N}{N} = k \, \Delta t.$$

Still another way to look at this is to rearrange the growth equation again, to say that the *fractional rate of change is constant*,

$$\frac{(\Delta N/N)}{\Delta t} = k.$$

In this equation the number of unicorns is discrete or quantized, that is, their number can only exist in terms of positive integers. Nevertheless, if we go to the limit of infinitesimal changes in number and infinitesimal time intervals, it can be shown by calculus that the population of the herd can be represented as a real exponential,

$$N(t) = N_o \, e^{kt} \ .$$

In this equation N_o is the number of unicorns when we started counting, that is, at time $t = 0$. In the very beginning there were only two unicorns, the Daddy and the Mommy. However, this relationship holds at any arbitrary time if N_o is the number at any starting time.

The number of unicorns at any time, $N(t)$, is a very sensitive function of k. A very small change in k will make a very large change in N, at any given time. If the unicorns are very amorous then k will be large.

Such a situation cannot go on forever; even if k is very small eventually the mass of the unicorn herd would be greater than the mass of Planet Earth. Long before this impossible state could occur food and water would have to be imported from other planets, but unicorns are said to restrict their diet to lotus blossoms and natural logarithms. Thomas Malthus (1766 – 1834) discussed this type of problem with respect to the human population and we are still wrestling with this problem today.

The reciprocal of k is often called the time constant. Some anthropologists, who have a way with words, call k the doom constant because it spells doom if k is positive *or* negative. It must be very close to zero to avoid disaster.

The important thing to note is that *all growth is transitory*. Growth cannot continue forever. In fact, growth only occurs for a short time, if you take the long view. What if humans kept growing all of their life-span like they grow in the first five years of life?

The time constant $1/k$ is often expressed as the *e*-folding time or the doubling time. By *e*-folding time we mean how long it takes for a number to be multiplied by e one time. Clearly this time, which we shall call T_e, is $1/k$, because,

$$N = N_o \, e^{k(1/k)} = N_o \, e^1 = 2.71828 \, N_o \,.$$

In terms of doubling time we would have

$$N = 2N_o,$$

so,

$$e^{kt} = 2.$$

This requires $\ln(e^{kt}) = \ln(2)$, so $kt = \ln(2)$ and therefore the doubling time, which we shall call T_2 is given by,

$$T_2 = \ln(2)/k = 0.693147/k = 0.693147 \, T_e \,.$$

Suppose $k = 0.1$ per year, which is like 10% compound ed interest. Then, in seven years,

$$N = N_o \exp[0.1 \times 7] = 2.01 \, N_o \,.$$

So the doubling time for $k = 0.1$ is almost seven years (actually about 6.93417 years).

If there were no wolves to eat the unicorns then the herd would double in about seven years; if there were no taxes to eat your investments then your assets would double in about seven years at 10% interest. It is interesting to compare exponential growth for 10% interest compounded annually and compounded continuously. Table 2-9 relates exponential growth for eight years, this is shown graphically in Fig. 2-8.

Table 2-9. Exponential Growth at 10% Compound Interest		
Time (years)	Total, Compounded Annually	Total, Compounded Continuously
0	1.0	1.0
1	1.1	1.1051709
2	1.21	1.2214028
3	1.331	1.3498588
4	1.4641	1.4918247
5	1.61051	1.6487213
6	1.771561	1.8221188
7	1.9487171	2.0137527
8	2.14358881	2.2255409

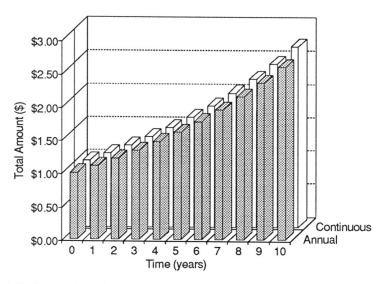

Fig. 2-8. Over a long interval, compounding continuously, quarterly, and semi-annually have big advantages over annual compounding. One dollar will double in approximately 7 years at 10% compound interest.

Even though both methods of compounding will double in about 7 years the continuous compounding soon increases much faster than the annual method. Some financial institutions offer daily compounding; this is better than annual, semi-annual, or quarterly compounding but not magnificently better for a short time.

In Fig. 2-9 we plot the unicorn population as a function of time on linear scales in both number and time. On this linear scale there is little indication for the first 30 or 40 centuries that there would be a problem of unicorn over-population. About 70 centuries a "population explosion" seems to occur but the problem was really there all along, as can be seen on the next page where the same data are plotted on a semi-log graph in Fig. 2-10.

It is said that using statistics is essentially a means of analyzing the past and projecting the future. Often a simple graph can indicate upcoming problems. This is clearly the case for the unicorn and human population. However, it is also clear that how you graph the data is important, too. The linear scale for exponential growth gives a false sense of security until the "explosion" occurs, whereas the log scale gives an early warning of impending disaster.

And now for the *bad* news. The situation is even more serious than I have stated (I didn't want to worry you). We started out with the rate of increase as,

$$\frac{\Delta N}{\Delta t} = kN.$$

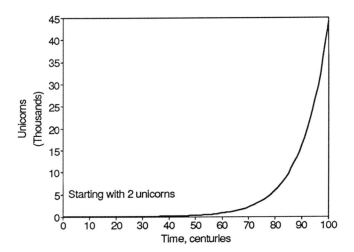

Fig. 2-10. Unicorn population viewed on linear scales. There does not seem to be a problem until about 70 centuries. k=0.1 per century.

This led to the number of unicorns growing exponentially,

$$N = N_o \, e^{kt} \; .$$

By substituting this expression for N into the right-hand side of the expression for $\Delta N / \Delta t$ we find that,

$$\frac{\Delta N}{\Delta t} = kN_o \, e^{kt} \; .$$

Now we see that not only is the *number* of unicorns increasing exponentially but the *rate of change* is increasing exponentially!

I hope this won't spoil your day but the situation is *even more serious*. The Facts of Life are that *all orders* of rates of change are increasing exponentially! This is because the exponential is invariant to differentiation, as our mathematician friends are fond of saying. This simply means that the *form* of the exponential is unchanged when we calculate its rate of change. Each rate of change is simply k times the previous one. For example, the rate of change of the rate of change (like an acceleration) is,

$$\text{rate of change of } \; \frac{\Delta N}{\Delta t} = \frac{\Delta}{\Delta t}\!\left(\frac{\Delta N}{\Delta t}\right) = \frac{\Delta^2 N}{\Delta t^2} = k^2 N_o \, e^{kt} \; .$$

Have a nice day.

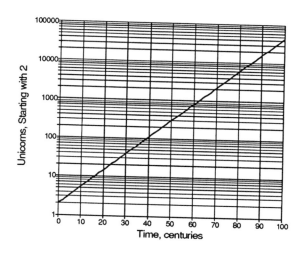

Fig. 2-11. Unicorn population viewed on a logarithmic scale, linear in time. It is clear that there was a problem at the beginning. $k = 0.1$ per century. Notice that you can easily see the original 2 unicorns on this scale.

Quiet Goes the Garbage

New York City's Sanitation Department has developed a quieter garbage truck which reduces the noise level to 67 decibels, compared to 85 decibels from present trucks.

New York State's anti-noise officials say that 80 decibels of sound imperils hearing. For comfortable living, they recommend a decibel limit of 45 at night and 65 by day.

Answers to Self-Tests

Answers to Self-Test 2-1	
(a) 13.979 or 14	(c) 6.3×10^{-4}
(b) −11	(d) 2.5×10^{-14}

Answers to Self-Test 2-2	
(a) 29.6	(c) 3.16×10^{-3}
(b) 26	(d) 4×10^{-2}

Postscript

Change is accelerating. Suppose we consider the time scale of known (written) human history as if it were the face of a clock. On this clock there were no significant changes in media in the first three thousand years. Nine minutes ago a revolutionary event occurred: The printing press was invented. During the last three minutes the telegraph, photograph, steam-powered ships and locomotives came into being. Two minutes ago we started using the telephone, the rotary press, motion pictures, the automobile, the airplane, and radio. One minute ago talking motion pictures were developed. In the last ten seconds television pervaded our lives. Five seconds ago computers appeared. Communication satellites and space flight came on the scene during the last second. The laser appeared about half a second ago followed by fiber optics in the last 30 ms.

Until about one minute ago there were almost no significant changes in medicine. According to Jerome Frank the whole history of medicine until then is the history of the placebo effect. About a minute ago antibiotics appeared. Ten seconds ago, open-heart surgery and organ transplants started. During the past two seconds there probably have been more changes in medicine than in all previous history.

There are more engineers and scientists alive now than have ever lived in the entire previous history of the world.

Digging Deeper

Waves and the Ear, W. A. van Bergeijk, J. R. Pierce, and E. E. David, Jr. (Doubleday and Co., Garden City, New York, 1960). Library of Congress Catalog Card Number 60-5948. This little paperback book by three authors at Bell Labs (biologist, physicist, and electrical engineer) is somewhat dated now in parts but it still makes fascinating reading.

The Speech Chain, Second Edition, P. B. Denes and E. N. Pinson (W. H. Freeman and Company, New York, 1993). ISBN 0-7167-2344-1. This is an easy introduction to human communication via speech and hearing by two electrical engineers at Bell Labs.

Chapter 3

More About Decibels

Parable of the Unicorns, Part II

It is enlightening to view the number of unicorns on a decibel scale, with time on a linear scale. To be consistent with our previous discussion of decibels we will define dBU (decibels unicorn) as

$$dBU = 10 \log \left(\frac{N}{N_o} \right)$$

where N is the number of unicorns at time t and N_o is the original number of unicorns, which we will use as the reference. It is common practice to use the value of a quantity at a particular time as a reference if no other is available (see Fig. 3-1).

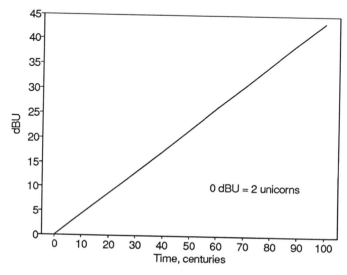

Fig. 3-1. Unicorn population viewed on a decibel scale, time on a linear scale.

Growth of Postal Rates

Things grow in several ways. Figure 3-2 shows how the postal rates in the United States of America have grown. One can extrapolate this curve and determine the year when the cost of mailing a letter will be equal to the entire Gross Domestic Product. It's not surprising that electronic mail is so popular.

Logistic Transition

Growth cannot continue forever. Some things go from one steady-state condition to another by an S-shaped path. For example, suppose that the herd consisted of a number N_o of unicorns that was constant because of a limited supply of Lotus blossoms, natural logarithms, and water. If a larger supply of food and water becomes available at time $t = 0$ the number of unicorns will increase and asymptotically approach a larger final number, N_f , by a path that depends on the particular processes involved.

   A common transition between steady states is by a "logistic curve" or "supply curve," which can be described by,

$$N(t) = \frac{A}{1 + B \exp[-kt]} .$$

In this equation A and B are constants such that the final number is $N_f = A$ and the original number is $N_o = A /(1 + B)$. This process is shown in Fig. 3-3.

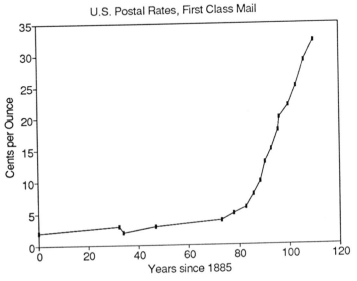

Fig. 3-2. Postal rates in the United States, as a function of time.

This expression is called the "logit" model by economists and those persons engaged in operations research for fun and profit. This is the well-known "S" curve or sigmoid. Nowadays there are not enough herds of unicorns to study, but experiments with populations of deer on an island and fruit flies in large containers bear this out, except that sometimes there are oscillations about the steady-state values of the populations that are not yet completely understood by people who study such things and who would be expected to understand them by now.

The logistic model is common because there is often a time delay between the cause and the start of a significant effect. The populations of unicorns and people cannot respond immediately to an increase in food and water because of the gestation times, 9 months for humans and 2.71828 years for unicorns. As the populations approach a new steady state or saturation point the increase decreases and slowly reaches the final state. This is different from an exponential process because an exponential process, like the charging and discharging of a capacitor through a resistor, starts immediately with a high rate of change and slows as the system approaches the final state.

Power Law Transition

Some processes occur without significant time delays, and the transitions may be represented by a simple power-law path, like the process shown in Fig. 3-4,

$$N(t) = 200 - \frac{100}{1 + t^n} \ .$$

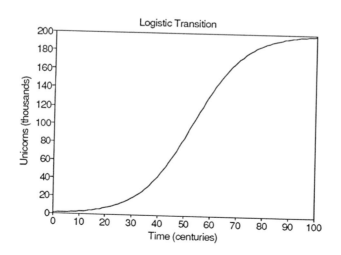

Fig. 3-3. Unicorn population increase due to an increase in available food and water. The logistic process is also called the supply curve.

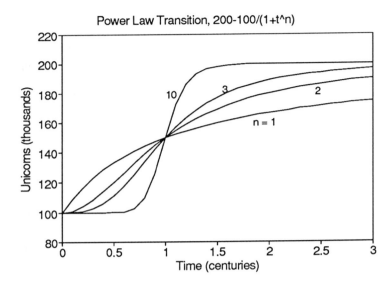

Fig. 3-4. Power-law transition. As n becomes large the power law begins to look like the logistic curve.

Exponential Transition

A simple exponential path is also a common process between steady state conditions. Like the power-law transition the exponential transition starts immediately and asymptotically approaches the new steady state, as described by,

$$N(t) = N_o(2 - e^{-kt}).$$

In this example, the unicorn herd starts with N_o unicorns and eventually the herd would double in size. The reciprocal of the time constant is k. The exponential transition is common in electronics, as in the charging and discharging of capacitors and inductors. Figure 3-5 shows an exponential transition of a unicorn herd to a new population state.

Temporal Decay. The Unicorn: Endangered Species?

Now let's examine a situation where the numbers are decreasing; perhaps the unicorns are losing interest in each other or perhaps people no longer believe in unicorns. Then,

$$N(t) = N_o e^{-kt}.$$

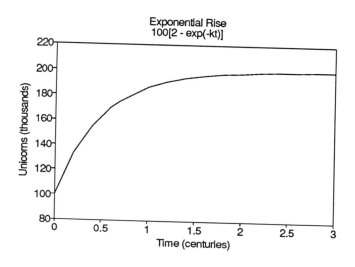

Fig. 3-5. Exponential transition between steady states. This exponential transition starts rapidly but continuously slows its rate of growth.

N dies away as shown in Fig. 3-6. It turns out that the universe is running down. The number of radioactive nuclei is decaying according to this equation. There were more nuclei before you started reading this; there will be less when you finish this sentence. The time constants for different nuclei vary from billionths of a second to billions of years but they're all going, on Planet Earth, on our Sun, and on the distant stars. Figs. 3-6 and 3-7 show typical results for counting radioactive decays.

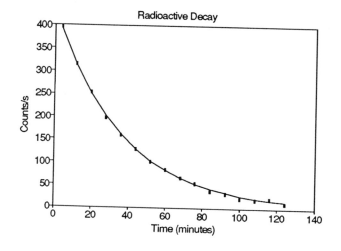

Fig. 3-6. Decay rate, linear scales. Statistics are better at high rates.

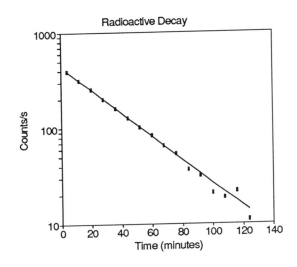

Fig. 3-7. These are the same data in Fig. 3-6, viewed on a semi-log scale.

If you want to watch the universe running down bit by bit, just look at a luminous watch dial with a magnifying glass. Each flash of light signals one less nucleus of that type. Nuclei are endangered species. Perhaps unicorns are so rare nowadays because their k is negative. The half-life is the time it takes for N_o things to decay to $N_o/2$. The relation between half-life and time constant is shown in Fig. 3-8.

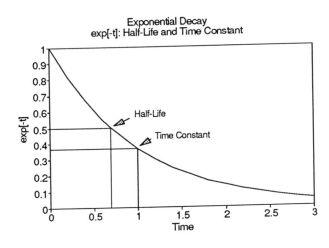

Fig. 3-8. Half-life and time constant for exp[$-t$]. Here the time constant recipro-cal, k, is 1 so the half-life is 0.693 and the decay time to $1/e$ is 1.

If you measure the temperature of your hot cup of coffee, decaffeinated or regular, and plot it as a function of time you will get an exponential decay to room temperature,

$$T(t) = T_r + (T_i - T_r)e^{-kt}$$

In the expression above, $T(t)$ is the temperature at any time, T_r is the ambient room temperature, T_i is the initial temperature when you started measuring, and k is the time constant which depends on the type of cup (ceramic or styrofoam) and its thickness and whether you put milk in your coffee. You can make k smaller by adding milk because the thin layer of fat on the surface reduces heat loss by evaporation. Here, too, you can watch the universe running down; the heat energy from your coffee can never be recovered completely and it was probably made by burning non-renewable fossil fuel. Even if you used solar energy to heat your coffee, that energy came from the sun which is running down. All of this is what people call the increase in entropy; we will see this word again in Chapter 11 when we talk about transmission of information. In Fig. 3-9 we see how sensitive the exponential is to k.

When sound energy is located in a closed space, it too runs down. Mostly, sound energy is absorbed by matter and converted to heat which can never be completely recovered.

When sound energy exists in a closed space and the energy source is suddenly cut off the sound intensity at a point in space decays approximately as,

$$I(t) = I_o e^{-kt}.$$

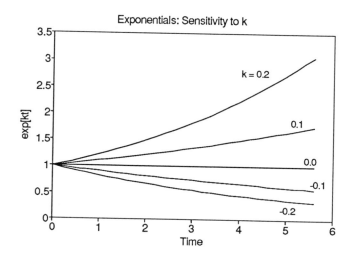

Fig. 3-9. Exponential growth and decay for various values of k.

Note: Here we are using the subscript "o" to indicate "original," not "output." In other words, we are using the original intensity as the reference, whatever it may be; the reference for dB(SIL) is not important here.

As with exponential growth we can inquire about the time constant for exponential decay. What is the *e*-folding time for intensity? Clearly, it is the same; $T_e = 1/k$, measured in seconds.

For exponential decay we often talk about the half-life, which is the time for the original number (the number when we started counting) to decay to half that number. We often designate the half-life as $T_{1/2}$. Clearly, the half-life is related to T_e by

$$T_{1/2} = \ln(2)/k = \ln(2)T_e \approx 0.693147\, T_e\,.$$

This is shown in Fig. 3-9.

Now, since intensity is proportional to the square of pressure, the time constant for pressure (or amplitude) decay, t_e, is related to that for intensity or energy by,

$$t_e = 2T_e = 2/k.$$

It is useful to express the temporal decay in decibels,

$$\frac{I(t)}{I_o} = e^{-kt}$$

$$\ln\frac{I(t)}{I_o} = -kt$$

In Chapter 2 we showed that $\log(x) = \ln(x)/\ln(10)$, so

$$dB = -4.34294\, kt.$$

The temporal growth and decay curves for several different values of k are shown in Fig. 3-9.

For sound in closed rooms the time constant $1/k$ must be restricted to a rather small range of values for speech intelligibility and enjoyment of music; all of these factors are contained in measurement and control of reverberation time. But, before that, we need to consider absorption and attenuation in more detail.

Absorption and Attenuation

When any wave, acoustic, electromagnetic, or other, is incident on a boundary there is the possibility of reflection by the boundary and transmission through the boundary. Consider, for example, light impinging on a glass window; some light is reflected and some is transmitted. There is usually some transfer of energy from the wave to the medium during reflection and transmission; this is what we call absorption. The energy loss from an acoustic wave to a medium can be attributed to several factors:

- Conduction loss or heat transfer.
- Viscous loss.
- Molecular energy loss.
- Wave interactions in solids.

Insertion loss is the loss in dB due to reflection and absorption. Insertion loss is 10 times the log of the ratio of incident power to transmitted power.

The Neper

Since natural logarithms appear so frequently in natural processes it would seem natural to define some unit like the decibel based on natural logs. This has been done, and it is used in some northern European countries as a telecommunications transmission unit. The neper is defined as,

$$N_p = 0.5 \ln(P_o/P_r)$$

where P_o is the output power and P_r is the reference power. If the impedance of the reference is the same as the output then we can express this in voltages,

$$N_p = \ln(V_o/V_r) \quad [V \text{ is in Volts}].$$

In this equation the voltages must be measured across the same value of impedance; that is, the impedances must be matched. See Fig. 3-11 for the effect of impedance mismatch on power transmission.

The decineper, or dN_p, is $0.1N_p$ but it appears not to be used as often as N_p. To convert decibels to nepers, multiply dB by 0.1151. Conversely, to convert nepers to dB, multiply N_p by 8.686. These units are very often used to describe attenuation per unit length in coaxial cable, fiber optic lightguide, and waveguide, in addition to bulk material. In this description the important parameter is the attenuation per unit length. For fused silica optical fiber, attenuation is a minimum of about 0.2 dB/km, while plastic optical fiber ranges from about 15 dB/km to 100 dB/km. Obviously, plastic fiber can only be used for short distances. Erbium-doped optical fiber can be used as a distributed amplifier. As an amplifier it will have a *negative attenuation* which is also expressed in nepers.

For purposes of comparison, Table 3-1 shows selected values of power ratio, impedance-matched voltage or current ratio, dB, and N_p. Some handy conversion factors are given in Table 3-2.

In Chapter 4 we will discuss reverberation, which depends in a fundamental way on energy loss in reflection, but first let's consider energy loss in transmission.

Table 3-1. Power, dB, and N_p			
Power Ratio	V or I Ratio (impedance matched)	dB	N_p
1.0233	1.0116	0.10	0.01
1.2589	1.1220	1.00	0.12
1.9953	1.4125	3.00	0.35
10	3.1623	10.0	1.15
100	10.0000	20.0	2.30
1 000	31.6228	30.0	3.45
10 000	100.0000	40.0	4.60

Table 3-2. dB to Neper Conversions		
Units	Multiply by . . .	. . . to obtain
dB/statute mile	7.154×10^{-2}	N_p/km
dB/nautical mile	6.215×10^{-2}	N_p/km
N_p/km	13.978	dB/statute mile
N_p/km	16.074	dB/nautical mile

Absorption in Transmission

When a plane wave propagates through matter we find that the intensity dies away in space due to absorption in accordance with the equation,

$$I(x) = I_o\, e^{-Kx} \ .$$

Here I is the intensity at x meters, I_o is the intensity at $x = 0$ meters (the start of the measurement), and K is the coefficient of power absorption which is often a function of frequency. For this reason K usually is measured at several frequencies or over a broad range of frequency. The unit of K is m^{-1}.

This expression is analogous to taking a photograph. At one point in time this equation describes a wide extent of space.

This exponential decay in space that is completely equivalent mathematically to an exponential decay in time. *Note:* We are using K for spatial decay and growth and k for temporal decay and growth. An example of the spatial decay of sound pressure is shown in Fig. 3-10.

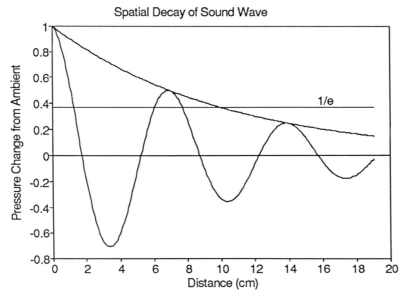

Fig. 3-12. Pressure at one instant in time, viewed over an extended interval of space. The pressure decreases as energy is transferred from the wave to the medium. Pressure change is in arbitrary units.

Absorption can be expressed in decibels as follows:

$$\frac{I_x}{I_o} = e^{-Kx}$$

$$\ln\left(\frac{I_x}{I_o}\right) = -Kx$$

In Chapter 2 we showed that

$$\log(x) = \frac{\ln(x)}{\ln(10)} = 0.434294\ln(x)$$

so,

$$dB = 10\log\left(\frac{I}{I_o}\right)$$

$$dB = -0.434294\,Kx$$

This is analogous to the expressions we obtained for temporal decay. Table 3-3 shows typical acoustic transmission losses of some common building materials.

| Table 3-3. Acoustic Transmission Loss for Selected Wall Materials |||||
Material	Thickness (inches)	Area Density (lb/ft^2)	Frequency (Hz)	Loss (dB)
Gypsum Wall Board	1.00	4.5	500	31
	2.00	9.0	500	34
	2.00	9.0	250	32
	2.00	9.0	1000	40
Glass	0.25	3.0	500	31
Concrete	4.00	53.5	500	45
Brick	12.00	121.0	500	53

In terms of decibels per meter,

$$\mathrm{dB}/m = -0.434294\,K.$$

Power Transmission Coefficient

This quantity is defined as the ratio of transmitted power to incident power. For sound it is often useful to describe the medium by the characteristic impedance R, measured in Rayls (in honor of Lord Rayleigh). In terms of the R values at the interface, the power transmission coefficient t_p is defined as

$$t_p = \frac{4\,R_1\,R_2}{(R_1 + R_2)^2}$$

R is the product of the volume density ρ (kg/m^3) times the speed of sound in the medium v (m/s).

Self-Test 3-1
A sound wave in medium #1 (with characteristic impedance 500 Rayls) is incident perpendicular to medium #2, characteristic impedance 1000 Rayls. What is the value of the power transmission coefficient?

Using a little algebra the power transmission coefficient can be expressed in a convenient form in terms of only one variable r, the *ratio of the R values* on both sides of the boundaries. This relation is graphed in Fig. 3-11.

$$t_p = \frac{4R_1/R_2}{(1 + R_1/R_2)^2} = \frac{4r}{(1 + r)^2}$$

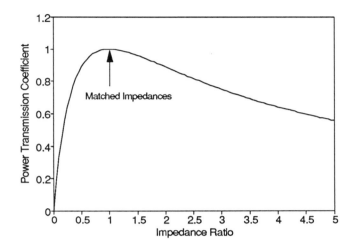

Fig. 3-11. Power transmission coefficient as a function of the ratio of characteristic impedances R_1/R_2. Maximum transmission occurs for matched impedances, $R_1 = R_2$.

In Fig. 3-11 you can see that power transfer is fairly tolerant of an impedance
mismatch in the neighborhood of a perfect match. A small mismatch will not make
much difference. This mismatch tolerance is much better on the high side of a perfect
match. The power transfer begins to fall off very fast as you move away from a
perfect match on the low side. Re-plot Fig. 3-11 with a log scale on the x-axis. You
may get a surprise.

Now you can distinguish two ways of isolating one system from another. One
way is by using absorbing media between them. A second way is to reduce the power
transmission coefficient by using a significant impedance mismatch at boundaries.
These two methods are usually effective over a wide band of frequencies. A third
method, which we won't go into here, involves adjusting impedances and thick-
nesses to reflect or transmit a narrow band of frequencies, like anti-reflection
coatings (filters) on optical surfaces. You can see this effect in the beautiful colors
reflected and transmitted by soap bubbles and oil films. These ideas are important
for stealth aircraft (radar) and submarines (sonar).

For solid media it is convenient to define the normal specific impedance,

$$Z_n = R_n + jX_n .$$

The specific acoustic impedance at any point is the ratio of the pressure phasor to
the velocity phasor. This is similar to the concept of impedance in electronics (see
Appendix 4, IWU #17 for a discussion of the symbol j). Appendix 5 shows how to
use phasors in analyzing simple filters.

Summary of Principal Points: Exponentials and Logs

As we have seen, exponential growth and decay in space and time are mathematically similar, only the physical interpretation is different. For ready reference let's summarize the principal points of exponential and logarithmic functions.

Exponential Decay			
Function	Half-Life	Time Constant (time to 1/e)	10% of Initial Value
$y = \exp[-kf]$	$t = 0.693/k$	$t = 1/k$	$t = 2.3/k$

Exponential Rise			
Function	Doubling Time	Time Constant (time to e)	10×Initial Value
$y = \exp[-kf]$	$t = 0.693/k$	$t = 1/k$	$t = 2.3/k$

$y = \ln (x)$						
x	0.1	1/e	0.5	1.0	2.0	e	10.0
y	−2.3	−1.00	−0.693	0.0	0.693	1.00	2.3

$y = \log (x)$						
x	0.01	0.1	0.5	1.0	2.0	10.0	100.0
y	−2.00	−1.00	−0.301	0.0	0.301	1	2

Exponentials Always Win in the End

Let's consider what really counts in our brief mortal span on Earth. Given enough time or range, nothing can beat an exponential. Over a short range geometric or power laws may seem to beat exponentials but all are eventually overtaken and surpassed. Figure 3-12 is worth 10^3 words.

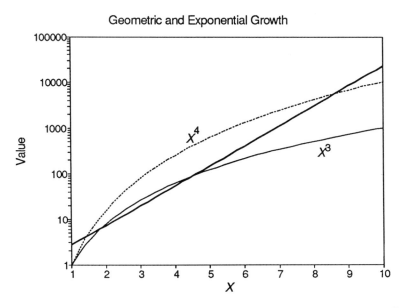

Fig. 3-12. The exponential always wins in the end. Thick line is e^x.

Table 3-8. Geometric and Exponential Growth			
x	x^3	x^4	e^x
1	1	1	2.7
2	8	16	7.4
3	27	81	20.1
4	64	256	54.6
5	125	375	148.4
6	216	1 296	403.4
7	343	2 401	1 096.6
8	512	4 096	2 980.9
9	729	6 561	8 103.1
10	1 000	10 000	22 026.5

You can bet on anything you want to but the exponential always wins in the end. So if the race is long, always put your money on the exponential. Obviously this advice also applies to negative powers and negative exponentials. The negative exponential eventually beats everything in the approach to zero. The exponential is relentless.

Answer to Self-Test

Self-Test 3-1
$$t_p = \frac{(4)(500)(1000)}{(500 + 1000)^2} = 0.89$$

Chapter 4

Reverberation and *Q*

What This Chapter is About

Now that you know how to handle the decibel in space and time we are ready to take up the subjects of reverberation and *Q*. *Q* will also introduce the use of the decibel in frequency. Reverberation refers to the gradual decay of energy as waves bounce around in an enclosed, or partially enclosed, space. *Q* is a measure, in a very general sense, of the ratio of energy stored to energy lost in a system.

- Reverberation is important in architectural acoustics, sonar, and electromagnetics. Acoustic reverberation has a profound effect on sound quality and intelligibility. For underwater acoustics the ocean can be thought of as a system with boundaries that reflect and absorb sound in ways which depend on temperature gradients and the type of boundary (water-to-air, water-to-ice, water-to-sand, etc.). Reverberation in electromagnetics occurs in things as small as a resonant cavity or a circuit board and as large as the space between the ionosphere and the Earth's surface.

- Reverberation time and *Q* are closely connected. *Q* is introduced in this chapter and is discussed in more detail in Chapter 5. The subject of *Q* naturally overlaps band-pass and band-stop filters, which are treated separately in Chapter 7 and Appendix 5.

Energy Loss in Reflection

Energy loss in reflection plays a central role in reverberation, which describes the build-up and decay of sound energy or electromagnetic energy in enclosed volumes. When a wave collides with a boundary some energy is reflected, some is absorbed, and some is transmitted. Only the reflected energy can contribute to *standing waves* because reflected waves combine with the incident waves to produce waves which appear to vibrate in place.

Standing Waves in an Enclosed Volume

When waves (any kind of waves) exist in an enclosed volume and the boundaries are good reflectors the stage is set for the phenomena of *resonance* and *standing waves*. By resonance we mean that certain frequencies will be selected by the system and the energy at these frequencies will be sustained longer than the energy at other, nonresonant, frequencies. Standing waves, which actually consist of traveling waves moving in opposite directions, occur at resonance and the resonant energy is carried by these waves. The energy at frequencies away from resonant frequencies dies away very rapidly so we can confine our attention to the resonant frequencies.

The number of dimensions of a system is a dominant factor in the systems characteristic spectrum. In one dimension you can think of a guitar or violin string where the system can only vibrate perpendicular to the string. In one dimension there will be only one set of characteristic modes of vibration. In two dimensions, think of a drum-head which can vibrate in two perpendicular directions. Now there are two sets of characteristic modes which may be independent or some modes which may be coupled. In three dimensions such as a rectangular room there are three independent, orthogonal modes which may be independent or possibly coupled. Mode coupling means that energy in one mode may be transferred to another mode. This becomes quite complicated and we will not discuss mode coupling.

To minimize complications let's start out with one dimension. Our discussion will be more like waves on a guitar string with one big difference: Sound waves are longitudinal waves (they wiggle in the direction in which they travel) but waves on a string, water waves, and electromagnetic waves are transverse waves (they wiggle perpendicular to the direction in which they travel).

The molecules of air are usually in random motion; a volume of gas is a disorderly system. When the wind blows or a sound wave propagates through the air there is a very small orderly motion superimposed on this disorder. When you consider a sound wave you can concentrate your attention on the particle displacement in the wave or, alternatively, the pressure in the wave. This tiny amount of orderliness in the displacement and pressure is what we perceive as sound information.

How tiny is it? At the threshold of human hearing the maximum displacement of an air molecule is approximately 10^{-11} m, which is about the size of a molecule. Atmospheric pressure is about 10^5 Pa, and at the threshold of human hearing the ear can detect a *change* in pressure of approximately 3×10^{-5} Pa. This is truly remarkable; the ear-brain system can detect a change of about 3 parts in 10^{10}. This is equivalent to measuring 3 nanovolts riding on a dc voltage of 10 volts!

At the upper limit of human hearing the pressure change from atmospheric is about 29 Pa, and the corresponding maximum displacement of air molecules is about 10^{-5} m.

At perfectly reflecting boundaries the particle displacement is zero because the particles cannot penetrate the boundary. Conversely, the pressure will have maxima at the boundaries because the particles will tend to be more dense there. Because the pressure is maximum where the displacement is zero we conclude that there is a phase difference of 90° between displacement and pressure.

When both boundaries are perfectly reflecting the displacement is zero at each boundary. Consequently any wave that is zero at both boundaries is a natural fit, or resonant wave. The simplest wave that will fit is a half-wave; in other words, the lowest natural frequency corresponds to a wavelength that is twice the distance between boundaries. The next frequency corresponds to a wavelength that is equal to the distance between boundaries, and so on. In general, any wave will fit if it has an integer number of half-waves equal to the distance between boundaries, and the resonant spectrum consists of evenly spaced frequencies, each one being an integer times the lowest resonant frequency.

WAVEMODE, Resonant Mode Worksheet

The worksheet named WAVEMODE is provided so you can experiment with standing waves viewed as displacement and pressure (or voltage). Retrieve this worksheet in the usual way and you should see a Home screen that looks something like Fig. 4-1.

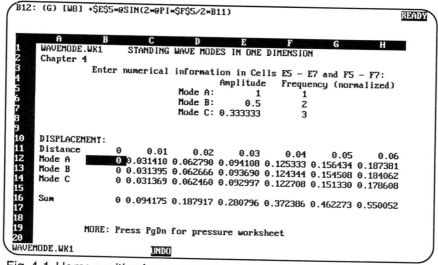

Fig. 4-1. Home position for WAVEMODE worksheet. You can view the corresponding screen for pressure or voltage by pressing the PgDn Key.

This worksheet lets you input your choice of modes and add them up to see the resultant at one instant in time. You can view the results graphically by selecting a named graph connected to this worksheet.

Worksheet Organization

Enter the amplitudes for modes 1, 2, and 3 in Cells E5, E6, and E7, respectively. It is useful to normalize the amplitudes to the lowest mode (amplitude =1) by entering the amplitudes as fractions. For example, type in the amplitude of mode 2 as 1/2 and the amplitude of mode 3 as 1/3. These appear as decimals in yourworksheet.

Enter the frequencies in Cells F5, F6, and F7. In this Figure the frequencies are normalized to one full wavelength. Therefore the lowest mode will be a halfwave and its normalized frequency is 0.5. Mode 3 is three halfwaves and is entered as 3/2. This appears in the worksheet as 1.5. Only three modes are shown but you can add as many as you like. Use the spreadsheet Copy command to copy the rows. Remember to modify the cell addresses in the formulas. Add the new modes to a graph using the Graph commands.

While viewing the Home screen of Fig. 4-1 press PgDn to see the corresponding screen for pressure standing wave modes. The graph named *Displace vs X* is shown in Fig. 4-2. It changes instantly as you make changes in the worksheet. Modes 1, 3, and 5 are shown in Fig. 4-3, and their amplitudes are 1/1, 1/3, 1/5, respectively. Observe the sum of the modes. Play around with the modes and you will see how Fourier series, and Fourier analysis and synthesis could have been discovered if Fourier had not done it in the early 1800's.

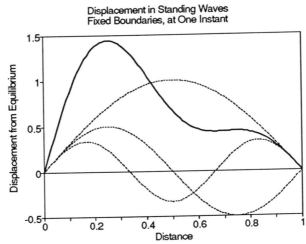

Fig. 4-2. The three lowest modes of displacement for a rectangular boundary with perfectly reflecting ends. Think of this as a photograph taken at one instant in time when all three modes are in phase.

Analysis decomposes a complicated wave into sines and cosines. Synthesis builds a wave with sines and cosines. Modes 1, 3, and 5 (for 1, 3, and 5 half-waves) are shown in Fig. 4-3.

MODEMOVE, Semi-Animated Worksheet

The WAVEMODE worksheet lets you experiment with user-defined resonant modes in one dimension. The results are like viewing a photograph at a single instant in time. With the MODEMOVE worksheet you can view selected modes at random instants in time so the results look like a group of photographs. Recalculate (usually function key F9) to see pulses ricochet from the boundaries. This is the essence of reverberation. The more modes that are excited, the more distinct is the pulse group, but you can see the effect with as few as three modes. Fig. 4-4 shows the Home screen for this worksheet.

Worksheet Organization

Cells E5, E6, and E7: Enter the amplitude of each mode in these cells.

Cells F5, F6, and F7: Enter the normalized frequency of each mode in these cells. The normalized frequency of the lowest mode is 1, third mode is 3, and so on.

Row 11 is the space axis. You can use the spreadsheet Fill command to change the increment and maximum range in this row.

Rows 12, 13, and 14 compute the displacement for selected modes. Place the cursor on a cell and its formula will be displayed at the upper left corner of the screen.

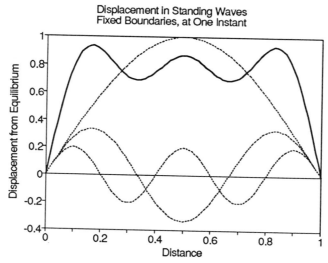

Fig. 4-3. Modes 1, 3, and 5 with amplitudes 1, 1/3, and 1/5 respectively, at one instant in time. Enter these data in the worksheet to see a graph similar to this.

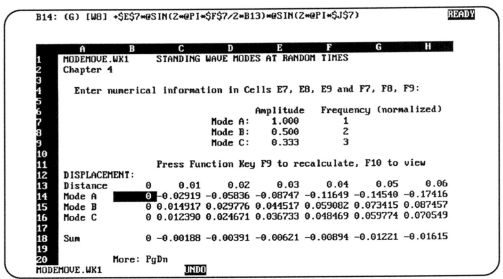

Fig. 4-4. Home screen of MODEMOVE . Cell B14 shows its formula at upper left.

Row 16 computes the sum of the three modes.

Press function key F10 to view the active graph. While viewing the graph press function key F9 (recalculate) twice, then press F10 again (*Lotus* commands). You will see the wave group at a different time, selected at random. See Fig. 4-5.

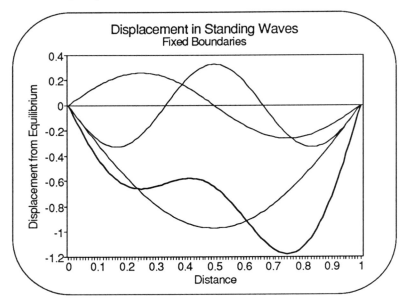

Fig. 4-5. Three modes and their sum at one instant in time. Press function key F9 to see what happens at another instant. Graph name: *Displace vs. X.*

Reverberation Time

Reverberation time is of critical importance in architectural acoustics because of its effect on speech intelligibility and music quality. Also, reverberation time in the human vocal tract is basic in speech production and is partially responsible for distinctive voice qualities, especially in vowels. It also plays an important role in sonar because of underwater reverberation. For the measurement of reverberation we position our measuring instruments at one or more points in space, and measure for a time interval.

The reverberation time of an enclosure is defined as the time for an initial sound intensity to fall 60 dB. This means that the intensity has decreased to 10^{-6} of its initial value. In terms of pressure, the sound pressure has decreased to 10^{-3} of its initial value at a time equal to the reverberation time. The decay is by an approximate exponential path.

Why 60 dB? Once again I'll tell you why: **Tradition!** About 1895, Professor W. C. Sabine of Harvard University made the first quantitative measurements of reverberation and he set the standard, as was his privilege. We now believe that the early decay time (defined as the time to –15 dB) is of primary importance. The early decay time is often abbreviated as EDT; this is also used for Eastern Daylight Time but context prevents confusion. The standard reverberation time is often abbreviated as RT_{60} or T_{60}. The reverberation time is the principal parameter for determining the acoustic quality of enclosed spaces.

The subject of reverberation is complicated because it is composed of reflections, absorption, phase shifts, and interference. These effects generally depend on frequency. If acoustic energy takes a long time to decay in an enclosure, then it will also require the same time to build up. In other words, when a sound is started in an enclosure it does not immediately reach full intensity but generally rises by an approximate exponential path. These time lags have a profound effect on voice and music quality. In addition, a long reverberation time generally reduces the rate at which information can be transmitted without error; the speaker must speak more slowly.

There have been three major advancements in the development of our current knowledge of reverberation:

- The Sabine equation

- The Norris-Eyring equation

- The Fitzroy equation

These equations and reverberation measurement techniques will be discussed in some detail as we proceed. Why do you always think that you sound better when you sing in the shower? There are two reasons. First, there is a long reverberation time due to highly reflecting walls but there are short time delays between reflections. Second, there is usually not a crowd of people to tell you how you really sound.

An idealized reverberation decay, consisting of a single decay with one time constant, is shown in Fig. 4-6.

The worksheet MODE-D-K is provided on the diskette so you can experiment with decays of single and multiple modes with time constants and amplitudes that you select.

MODE-D-K, Multi-Mode Reverberation Worksheet

This worksheet lets you experiment with single-mode and multi-mode decays. You view the results in one point in space as the modes decay in time. Remember that the total pressure at an instant is the sum of the individual pressures. Here we are using the instantaneous pressure, *not* the rms pressure. Fig. 4-7 shows the Home screen of the worksheet. When you view this on your monitor your can press the PgDn Key to view the second screen.

Worksheet Organization

Cells D5 and D6: Enter your choice of amplitudes for the modes.
Cells E5 and E6: Enter your choice of frequencies (Hz) for the modes.
Cells F5 and F6: Enter your choice of time constants (s) for the modes.
Row 9: This is the time axis. You can change the time increments by means of the spreadsheet Fill command.

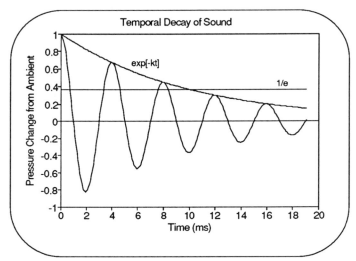

Fig. 4-6. Idealized decay of sound pressure in a single mode. Usually measurements involve several modes, each with a different time constant. This graph is for one point in space, over an interval of time. Compare Fig. 3-12.

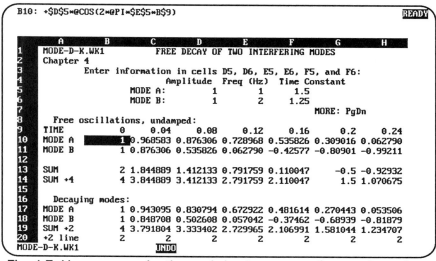

Fig. 4-7. Home screen for the worksheet MODE-D-K. Cursor shows formula in Cell B10 at upper left of screen. It's easy to include more modes in MODE-D-K.

Row 10: This row contains the undamped time evolution of Mode A.

Row 11: This row contains the undamped time evolution of Mode B.

Row 13: This row contains the sum of Rows 10 and 11, the undamped modes.

Row 14: For graphical clarity this row adds +4 to the values of Row 13, to graph the sum in aspace above the individual modes.

Row 17: This row contains the damped time evolution of Mode A.

Row 18: This row contains the damped time evolution of Mode B.

Row 19: This row contains the sum of Rows 17 and 18,with +2 added for graphing.

Row 20: This row contains a horizontal line at +2 to provide an axis at that level. Press the PgDn key to see the following rows:

Row 21: The total energy of both modes is calculated in this row.

Row 22: The sum of both modes is calculated in dB in this row.

Row 23: This calculates the exponential decay for the time constant of Mode A.

Row 24: This calculates the exponential decay for the time constant of Mode B.

Row 26: This row calculates the exponential decay of energy for Mode A.

Row 27: This row calculates the exponential decay of energy for Mode B.

To view the undamped modes and their sum select the graph named *Undamped*. You should see the graph shown in Fig. 4-8. Experiment with different amplitudes and frequencies.

Next select the graph named *Damped* and you should see the graph shown in Fig. 4-9. Now you can experiment with the time constants as well as the amplitudes and frequencies. Observed the sensitivity to the choice of time constants.

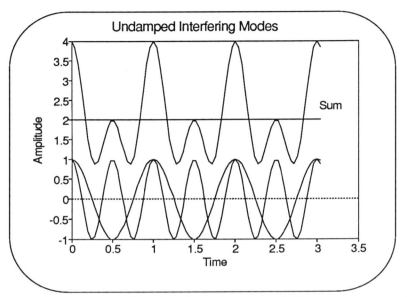

Fig. 4-8. Two undamped modes viewed at one point in space, over an interval of time. Top: Superposition (sum) of modes. Bottom: Individual modes.

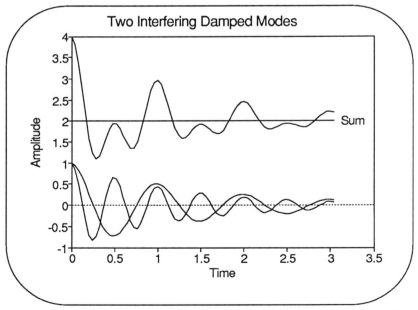

Fig. 4-9. Decay of two modes viewed at one point in space, over an interval of time. Sum is shown at the upper trace. Graph name: *Damped*.

Select the graph named *dB* and you should see the graph shown in Fig. 4-10. Notice the interference in the decay because of the two modes, Fig. 4-11. The envelopes of both time constants are shown. Set the amplitude of one mode to zero and observe the envelope of the decay. An idealized decay in shown in Fig. 4-12.

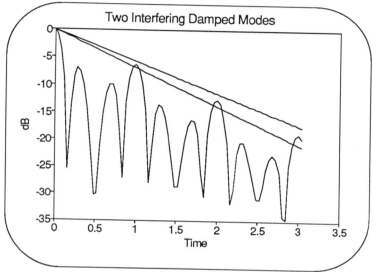

Fig. 4-10. Decay of two modes with different time constants. Graph name: *dB*. The two straight lines are the envelopes of the individual modes.

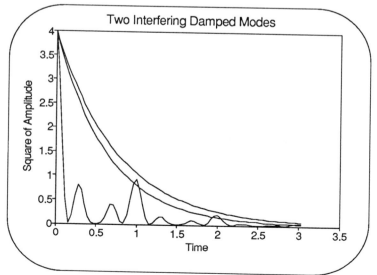

Fig. 4-11. Decay of power for two interfering damped modes.

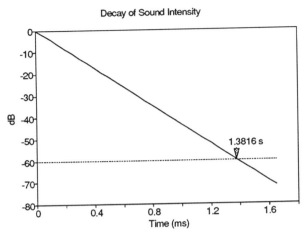

Fig. 4-12. Decay of sound intensity in dB for a typical theatre. This is an idealized decay curve, for a single mode.

There are difficulties with multiple modes in most rooms and auditoriums; the decay is not a simple exponential envelope because many decay times are superimposed and different frequencies have different decay times. The situation can be described completely by the cryptic statement that the Qs of the modes are different (the meaning of this will be revealed soon). The energy in the lower-Q modes will die out fastest, just as the short-lived nuclei die out fastest. Nevertheless, one can still measure the time interval for the intensity to fall 60 dB even though it will not be a simple, single exponential decay.

Although speech intelligibility will be high, a very small reverberation time makes a room sound muffled and acoustically dead. Most speakers prefer a little reverberation. On the other hand, a long reverberation time makes a room sound hollow and can interfere with speech intelligibility and the enjoyment of music because of long-delayed multiple echos. On the other hand, a long reverberation time may be desirable in a church where the echoes are psychologically associated with authority.

A long reverberation time can cause a human to speak with difficulty because of the altered feedback delay. When we speak we receive feedback through bone conduction and air conduction, and we are conditioned to speak normally in accordance with these two time delays. As with any servomechanism or feedback control system, a change in feedback delay can have serious consequences for stability in a system that is otherwise functioning normally. Some forms of stuttering have been attributed to altered feedback delay, and stock market gyrations due to high-speed computer program trading and the elimination of the 6-month capital gains holding time are examples of this problem.

Speech perception by someone other than the speaker is also affected by reverberation time. As the sound pattern of speech changes in amplitude and frequency the longer build-up and decay associated with long reverberation times makes older sounds overlap with newer sounds resulting in confusion and reduced intelligibility; previous syllables will partially mask later ones.

In telecommunications, reverberation is associated with multipath interference of electromagnetic waves. You can see one of these effects when you are watching TV or listening to an FM radio station and an aircraft flies over. The ghost images on the TV screen and the wobbling sound on the radio are due to interference between the direct wave from the transmitter and the reflected wave from the aircraft. The space in the spherical shell between the surface of Planet Earth and it's ionosphere can be observed to reverberate due to lightning strokes.

Table 4-1 gives approximate values for what is thought to be optimum reverberation times for enclosures arranged according to use.

Table 4-1. Preferred Reverberation Times		
Reverberation Time (seconds to −60 dB level)	Room Use	Room Volume (cubic meters)
0.4 - 1.1	Theaters (voice)	30 - 7,000
0.4 - 1.6	Movie theaters	170 - 60,000
0.9 - 2.6	Dance music	250 - 25,000
0.9 - 3.2	Light music	70 - 50,000
1.4 - 2.9	Concert studio	90 - 70,000
1.8 - 3.2	Orchestral music	300 - 40,000
2.1 - 3.9	Church music	600 - 70,000

Reverberation time has a profound effect on speech intelligibility for two reasons. First, the delayed feedback can confuse the speaker. Second, multiple echoes can confuse the listeners.

RT-60, Reverberation Time Worksheet

Retrieve the worksheet RT-60 in the usual way and you should see the Home screen in Fig. 4-13. If you do not see this press the Home key.

This worksheet lets you input a time constant and view the rms intensity of a single mode as it decays. The worksheet also calculates and displays the reverberation time, the time to −60 dB corresponding to the selected time constant. (Recall that the time constant is the time for decay to the $1/e$ point.)

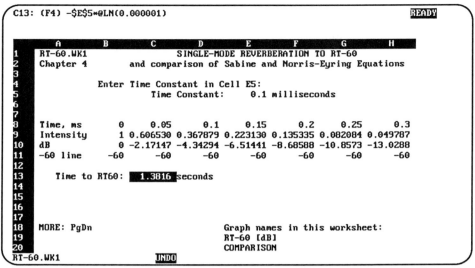

Fig. 4-13. Home screen for RT-60 worksheet. The cursor highlights the formula in cell C12 which shows the reverberation time corresponding to the selected time constant. Press PgDn for Sabine and Norris-Eyring comparison.

Worksheet Organization

Cell E4: Enter time constant in milliseconds. You can change the units of the time constant but be sure to keep Row 7 in the same units of time.

Row 7: This is the time axis. Use the Fill command to adjust the increment.

Row 8: This is the rms intensity.

Row 9: This row calculates the signal in dB.

Row 10: This provides a horizontal line at -60 dB.

Cell C12: This cell displays the time to RT_{60}.

The worksheet graph connected to the worksheet is shown in Fig. 4-14.

Sabine Equation

The Sabine equation, often used as a quick approximation for the reverberation time when the absorption is low, can be expressed in metric units as

$$T = 0.161 \frac{V}{\alpha}$$

where T is the time in seconds to the -60 dB level, V is the enclosed volume in cubic meters, and α is the average absorption which is the sum of all the surface areas multiplied by their respective absorption coefficients, all divided by the total area.

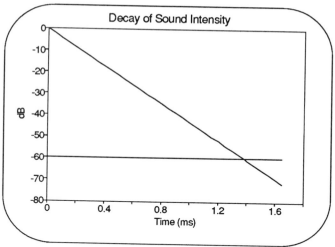

Fig. 4-14. Worksheet graph for an ideal decay. Graph name: *RT-60 [dB]*.

In common engineering units of feet and seconds, the Sabine equation has the form

$$RT_{60} = 0.049 \frac{V}{\alpha}$$

where V is the enclosed volume in cubic feet and α is the absorption expressed in *sabins*. The sabin is an absorption unit which is characteristic of a surface that absorbs sound at the same rate as one square foot of a perfectly absorbing surface or an open window (see Digging Deeper section).

For example, 10 ft^2 with an absorption coefficient of 0.1 would be the same as 1 ft^2 of a perfect absorber, so the 10 ft^2 would be equivalent to one sabin. Similarly, 10 m^2 with an absorption coefficient of 0.1 would be the same as 1 m^2 of a perfect absorber, so it would be equivalent to one metric sabin. Another common form of the Sabine equation in SI or mks units is

$$RT_{60} = 0.161 \frac{V}{S\alpha_{av}}$$

where S is the total interior surface in m^2 and α_{av} is the average absorption coefficient of the surfaces. If a source of continuous sound exists in an absorbing enclosure for a long time eventually a steady-state energy density will be established. The energy input will equal the energy absorbed. The steady-state intensity I_f is related to the power W and the absorption α by,

$$I_f = \frac{W}{\alpha} \quad \text{Watts/m}^2$$

Enclosures with small absorption (high Q) can store large amounts of energy. The Sabine equation is simple and easy to use but it cannot be correct if the absorption is large. Consider the case in which the absorption coefficients of all surfaces are 1, so that there is no reflection. Physically, the reverberation time must be zero but the Sabine equation predicts $T = 0.161V/S$ for this case. In fact, if the average absorption coefficient is as high as 0.2 then the Sabine equation is in error by about 10%.

The difficulty with Sabine's approach is that it assumes reflections produce a *uniform energy density distribution* throughout the volume. In the worksheets WAVEMODE and MODEMOVE we have seen that the energy distribution is far from uniform when only a few modes are present. When Professor Sabine made his measurements he used reflecting planes near the center of the rooms to mix up the wave modes and produce a more uniform distribution, but this only works if the sounds persist long enough for the mixing to occur. That is why the Sabine equation is restricted to fairly long reverberation times, that is, small absorption.

Norris-Eyring Equation

Norris and Eyring were worried about the Sabine equation, which predicted a finite reverberation time with 100% absorption. They also showed that the Sabine equation gave values greater than 100% for the average absorption coefficient when the true absorption coefficient was greater than 0.63. To correct these difficulties they took into account multiple reflections and derived the following relationship,

$$RT_{60} = \frac{0.049 \, V}{-S \ln(1 - \alpha_{av})} \, .$$

In this equation British engineering units are used. This equation gives values for α_{av} from 1.0 to 0 for true absorption calculated from actual RT_{60} measurements.

A comparison of the Sabine and Norris-Eyring equations shown in Fig. 4-15. Agreement is good for small absorption, as expected.

One of the most common methods of measuring absorption coefficients is to measure RT_{60}. When using tables of absorption coefficients it is extremely important to know whether the Sabine equation or the Norris-Eyring equation was used in their calculation from measurements.

Fitzroy Equation

The Norris-Eyring equation assumes that the absorbing material is evenly distributed and gives good results in such a situation. A more common situation in real life is that the absorbing material is not evenly distributed.

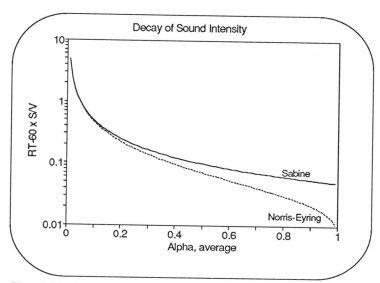

Fig. 4-15. Comparison of Sabine and Norris-Eyring equations.

For example, the ceiling may be covered with acoustic tile but the floor may be wood, marble, carpet, etc., and the walls may be highly reflective. The Fitzroy equation reduces to the Norris-Eyring equation for evenly distributed absorbing material. In all its glory, the Fitzroy equation in British engineering units is,

$$RT_{60} = 0.049 \frac{V}{S^2} \left[\frac{2xy}{-\ln(1-\alpha_{xy})} + \frac{2xz}{-\ln(1-\alpha_{xz})} + \frac{2yz}{-\ln(1-\alpha_{yz})} \right].$$

Here x, y, and z are the height, width and length of the room, and α_{xy} is the average absorption coefficient of the two end walls, α_{xz} is the average absorption coefficient of the two side walls, and α_{yz} is the average absorption coefficient of the floor and ceiling. Typical absorption coefficients of some surfaces are shown in Table 4-2.

Table 4-2. Absorption Coefficient α						
	125 Hz	250 Hz	500 Hz	1000 Hz	2000 Hz	4000 Hz
Plywood 1 cm	0.28	0.22	0.17	0.09	0.10	0.11
Acoustic tile 2 cm	0.76	0.93	0.83	0.99	0.99	0.94
Glass window	0.35	0.25	0.18	0.12	0.07	0.04
Gypsum board	0.29	0.10	0.05	0.04	0.07	0.09
Heavy carpet	0.02	0.06	0.14	0.37	0.66	0.65

The Fitzroy equation can be used to calculate RT_{60} from an architect's plans. This value is then used in the Norris-Eyring equation to obtain the value of α_{av}.

Measurement of Reverberation Time

Measurement of reverberation time has a twofold purpose. First, knowledge of reverberation time enables one to fine-tune the acoustics to the purpose of the room. Second, it is very common to use reverberation time measurements in chambers lined with acoustic material in order to measure the absorption coefficient of the material.

Reverberation time can be estimated with equipment as simple as a sound level meter and a stopwatch. Modern instrumentation is available for more accurate measurement. Figure 4-16 shows the use of an impulsive sound source (a pistol shot) used to measure reverberation time by means of a digital sound level meter. The Fourier transform of the sound decay gives the room's frequency response (see Chapter 9).

Fig. 4-16. Engineer with pistol and ear protectors measures reverberation time with a sound level meter. Caution: Use blank cartridges. This sound level meter can store data and download to a computer. (Courtesy Brüel & Kjaer Instruments, Decatur, GA)

Digging Deeper

The Physics of Sound, R. E. Berg and D. G. Stork (Prentice-Hall, Englewood Cliffs, New Jersey, 1982). ISBN 0-13-674283-1. This is an introduction to the physical basis of acoustics and music for nonscientists. It has very little mathematics.

Fundamentals of Acoustics, Third Edition, L. E. Kinsler, A. R. Frey, A. B. Coppens, and J. V. Sanders (John Wiley & Sons, New York, 1982). ISBN 0-471-02933-5. For many years this book has set the standard. It treats acoustics in mathematical detail and some parts are not easy. Chapter 13 discusses architectural acoustics.

Acoustics: An Introduction to Its Physical Principles and Applications, A. D. Pierce (McGraw-Hill Book Company, New York, 1981). ISBN 0-07-049961-6. This book is directed to upper-level and graduate students in engineering, physics, and mathematics. It is somewhat more theoretical than Kinsler *et al.* and has less coverage of sonar and underwater acoustics. Chapter 6 discusses room acoustics.

A Dictionary of Scientific Units Including Dimensionless Numbers and Scales, H. G. Jerrard and D. B. McNeill (Chapman and Hall, London, 1980), page 99. In 1911 Professor Wallace C. Sabine named the unit of sound absorption the "open window unit" or o.w.u. It is equal to the sound absorption by an open window of 1 square foot in area. In 1937 this unit was renamed the sabin. Professor Sabine performed most of his measurements at 512 Hz and for a long time this was the traditional frequency used in absorption measurements. Gradually the standard changed to 500 Hz. Now it is realized that measurements should be conducted over a wide band of frequencies.

There is a natural rock cistern in Ft. Warden, WA, that is 186 feet (56.7 m) deep. It's acoustics are dominated by remarkable reverberation properties; the local citizens call it "the Cistern Chapel."

Chapter 5

More About the Decibel and *Q*

What This Chapter Is About

The importance of Q is that it is a measure of the ratio of energy stored to energy lost in a system, per half cycle. You have seen in Chapter 4 that reverberation time depends on energy storage and loss properties of a system. The concept of Q is so pervasive throughout engineering and the sciences because it describes these processes in very general terms, independent of the particular system.

Table 5-1 shows the analogies between three types of systems. Energy is stored in inductance and capacitance and their analogous quantities, and energy is transformed into thermal form in resistance. Q can be expressed in terms of these quantities and it provides a general description of system behavior in both the time and frequency domains. Worksheets are provided so you can experiment with Q.

Table 5-1. Analogous Elements in Electrical, Mechanical and Acoustic Systems					
Electrical		Mechanical		Acoustic	
Element	Symbol	Element	Symbol	Element	Symbol
Voltage	V	Force	F	Pressure	P
Charge	q	Displacement	x	Volume displacement	X
Current	I	Velocity	v	Volume current	V
Capacitance	C	Compliance	$1/k_m$	Capacitance	C_a
Inductance	L	Mass	m	Inertance	M
Resistance	R	Resistance	R_m	Resistance	R_a

The Decibel and *Q*

Before people understood all about filters there were those who thought that the quality of a filter (wharerever that is) could be measured by its bandwidth or selectivity, and the more narrow the bandwidth the higher the quality. Perhaps this was because of the difficulty of obtaining very narrow bandwidths in the early days of filters.

Loaded *Q* and Unloaded *Q*

In real life we always measure "loaded" values of *Q*, which take into account the finite impedances of the source and the measuring instrument (or actual load). Nevertheless, when the source and load impedances are much larger than the impedance of the system under test then they can be neglected and it is only the parameters of the system under test that determine the "unloaded" *Q*. Loading always lowers *Q* but measurement at the –3 dB points is valid for both loaded and unloaded *Q*. Figures 5-1 and 5-2 define *Q* graphically.

Q Measurement at –3 dB Points

The –3 dB point is arbitrarily called the cut-off frequency of low-pass and high-pass filters. Also, the frequency interval between the –3 dB points of a band-pass or band-stop filter is called the bandwidth Δ*f* (the bandwidth of hearing aids is defined differently; see Appendix 3).

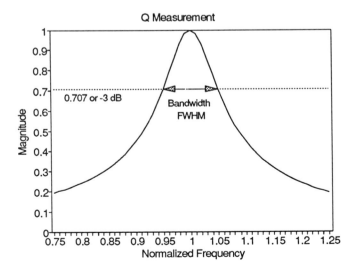

Fig. 5-1. *Q* measurement by $f_r/\Delta f$ method. FWHM=Full Width Half Maximum. Half-maximum power is 0.707 maximum voltage or displacement.

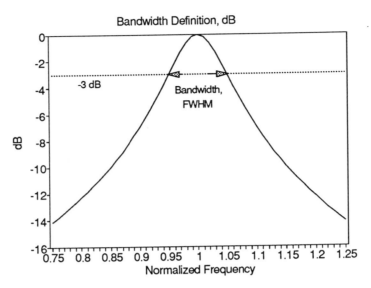

Fig. 5-2. Q measurement by bandwidth between the –3 dB points on the power resonance curve (full-width half-maximum power). In this Figure, $Q = 10$.

The symbol Q is defined as the reciprocal of the relative bandwidth,

$$Q = \frac{f_r}{\Delta f} \; .$$

In this definition of Q, f_r is the resonant frequency (where the system response is its maximum) and Δf is the bandwidth. This definition is consistent with the definition of Q as the ratio of energy stored to the energy lost, per half cycle.

Examples

(a) The lower half-power point ($0.707 \times$ maximum) corresponds to a normalized frequency of 0.99, and the upper half-power point corresponds to a normalized frequency of 1.01. The normalized resonant frequency is 1.0. This results in a Q given by,

$$Q = \frac{1}{\Delta f} = \frac{1}{1.01 - 0.99} = 50.$$

(b) The lower half-power point (–3 dB) corresponds to a normalized frequency 0.95 and the upper half-power point corresponds to a normalized frequency of 1.05, so the Q is 10. Now try Self-Test 5-1.

Self-Test 5-1

The resonant frequency is 10 kHz.

(a) For a *Q* of 57, what is the bandwidth (Hz) ?
(b) What is the frequency of the lower −3 dB point (Hz) ?
(c) What is the frequency of the upper −3 dB point (Hz) ?

Q Measured at Other dB Points

The standard measure of *Q* in the frequency domain is based on the −3 dB points, but suppose you need to calculate *Q* based some other attenuation level. Fortunately the method is simple. Let V_r be the voltage of the center frequency f_r and let V be the voltage associated with some other full-width frequency interval Δf, symmetric with the center frequency. Now *Q* is given by,

$$Q = \frac{f_r}{\Delta f}\left[\left(\frac{V_r}{V}\right)^2 - 1\right]^{1/2}.$$

Remember that here Δf is *not* the bandwidth except for the special case of $V_r/V = 1.414$, which corresponds to the −3 dB level.

Examples

(a) Suppose f_r is 10 kHz and Δf is 100 Hz when V_r/V is 1.122. (This is the −1 dB level.). These data give $Q = 50.88$.

(b) Suppose f_r is 10 kHz and Δf is 196.5 Hz when V_r/V is 1.414. (This is the usual −3 dB level.). These data give $Q = 50.88$.

Self-Test 5-2

At the resonant frequency, 1 kHz, you measure 1 Volt. At 950 Hz and at 1050 Hz you measure 0.5 Volt. Compute the *Q*.

Shape Factor and Skirt Selectivity

The *shape factor* is defined as the ratio of the −60 dB bandwidth to the −3 dB bandwidth. For example, if the −60 dB bandwidth is 15 kHz and the −3 dB bandwidth is 3 kHz, then the shape factor is 5. The *skirt selectivity* refers to the rate of increase in attenuation as you move away from the −3 dB points. A small shape factor is associated with good skirt selectivity.

Q Measurement by Rate of Change of Phase

In the frequency domain *Q* can also be measured by the *rate of change of phase* between the −3 dB points,

$$Q = \frac{\omega_r}{2} \left| \frac{d\theta}{d\omega} \right| .$$

In this equation ω and θ are measured in radians/s and radians, respectively. This can also be expressed with the frequency in Hz and θ in radians,

$$Q = \frac{f_r}{2} \left| \frac{d\theta}{df} \right| .$$

The method of measurement is shown in Fig. 5-3. The quantity $-d\theta/d\omega$ is called the *group delay*. The group delay is useful when the power resonance is broad because the phase is a rapidly varying function when the magnitude is slowly changing.

This expression for Q also carries some important physical insight. This shows that the higher the Q, the longer the signal is trapped in the resonant system.

Example
If the phase is +45° at a normalized frequency of 0.99 (the lower –3 dB point) and the phase is –45° at a normalized frequency of 1.01 (the upper –3 dB point), then

$$Q = \frac{1}{2} \left(\frac{\pi/4 - (-\pi/4)}{1.01 - 0.99} \right) = 39.27.$$

This underestimates Q because of the curvature near the –3 dB points. Measuring the phase closer to the resonant frequency gives a more accurate result, close to 50.

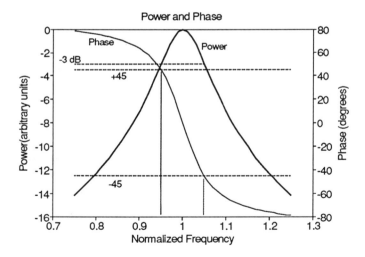

Fig. 5-3. Q measurement by rate of change of phase with respect to frequency (group delay).

Measurement of *Q* in the Time Domain

The *Q* of a resonant system can also be measured, consistent with this definiton, in the time domain by the *logarithmic decrement*. The logarithmic decrement *d* is the difference of the natural logs of the amplitudes on any two adjacent swings (see Fig. 5-4). By the rules of logarithms this is equivalent to $\ln(A_n/A_{n+1})$. In terms of *d* the *Q* value is given by,

$$ Q = \frac{\pi}{\ln(d)} = \frac{\pi}{\ln(A_n/A_{n+1})} \ . $$

Example
If the ratio of adjacent maxima is 0.699/0.488 then the *Q* is,

$$ Q = \frac{\pi}{\ln(d)} = \frac{\pi}{\ln(0.699/0.488)} = 8.7 $$

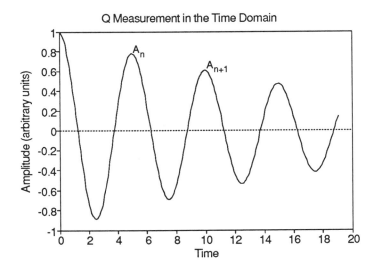

Fig. 5-4. *Q* measurement by logarithmic decrement in the time domain. This method applies to voltage measurements as well as pressure measurements.

Self-Test 5-4

Suppose you are measuring a decaying sine wave. You find a maximum is 9.99 *V* and the next maximum is 9.00 *V.* What is the *Q* ?

Unloaded *Q* of *LCR* in Terms of *LCR*

For a circuit composed of *L*, *C*, and *R* (see Chapter 7 and Appendix 5) the unloaded *Q* is given as $2\pi f_r L/R$, and the resonant frequency f_r is given by

$$f_r = \frac{1}{2\pi\sqrt{LC}} \; .$$

Combining these relations we find several equivalent ways of expressing *Q*,

$$Q = \frac{(L/C)^{1/2}}{R} = \frac{2\pi f_r L}{R} = \frac{\omega_r L}{R} \; .$$

This shows explicitly how the unloaded *Q* depends on *R* and the *L/C* ratio. So, for fixed *R*, the *LC* product controls the resonant frequency but the *L/C* ratio controls the *Q*. In other words, many values of *L* and *C* can be chosen for a particular resonant frequency but some combinations may have undesirably large or small *Q*s.

*Q*s of Some Physical Systems

The range of *Q*s of physical systems is enormous. There are few physical quantities with a range as wide as 10^{26}. Some representative *Q*s are shown in Table 5-2. The Mössbauer effect has a very large *Q* because it involves very long-lived (lightly damped) energy levels in nuclei of iron and iridium isotopes.

Table 5-2. *Q*s of Some Systems	
System	*Q*
LCR circuit	$1\!-\!10^3$
Pendulum	$1\!-\!10^5$
Microwave resonant cavity	$10^4\!-\!10^5$
Quartz crystal (selected)	$10^6\!-\!10^7$
Laser, maser	10^{14}
Superconducting cavity	10^{15}
Mössbauer effect ($_{26}Fe^{57}$, $_{77}Ir^{191}$)	$10^{11}\!-\!10^{26}?$

Rule of Thumb for *Q*

If the amplitude of a maximum is about 11% greater than the next maximum, then the Q is about 30.

Summary of *Q* Relations

Q in terms of the frequency difference between half-power points:

$$Q = \frac{\omega_r}{\Delta\omega} = \frac{f_r}{\Delta f} \; .$$

Q in terms of the rate of change of phase (in radians) near resonance:

$$Q = \frac{\omega_r}{2} \left| \frac{d\theta}{d\omega} \right| = \frac{f_r}{2} \left| \frac{d\theta}{df} \right| \; .$$

Q in terms of the logarithmic decrement d :

$$Q = \frac{\pi}{\ln(d)} \; .$$

Q in terms of relaxation time t_e for amplitude decay to $1/e$ of starting value:

$$Q = \omega_r \frac{t_e}{2} = 2\pi f_r \frac{t_e}{2} \; .$$

Q in terms of relaxation time t_h for amplitude decay to $1/2$ of starting value:

$$Q = \omega_r \frac{t_h}{2 \ln(2)} = 2\pi f_r \frac{t_h}{2 \ln(2)} \; .$$

ABC-OF-Q, Frequency and Time Worksheet

The worksheet ABC-OF-Q lets you experiment with Q in the frequency and time domains. Start your spreadsheet and retrieve this worksheet in the usual way. You should see the Home screen shown in Fig. 5-5. If you do not see this press the Home key. The time-domain screen below Row 20 (not shown here) can be viewed on your monitor by pressing the PgDn key.

In this worksheet the resonant frequency is normalized to 1. All you have to do is to enter the time constant t_e in Cell E3. The frequency response is then computed in terms of the magnitude, power, dB, phase (rad) and phase (deg). The amplitude for free decay in the time domain is also computed. All of these are available as named graphs that you can view in the usual way.

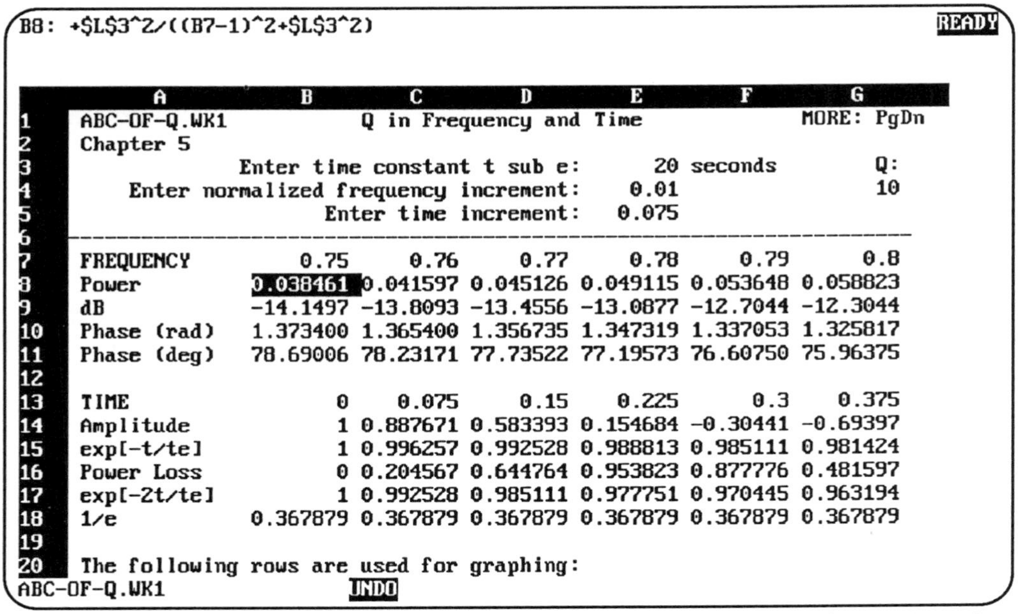

```
B8: +$L$3^2/((B7-1)^2+$L$3^2)                                          READY

        A           B        C        D        E        F        G
1  ABC-OF-Q.WK1              Q in Frequency and Time           MORE: PgDn
2  Chapter 5
3              Enter time constant t sub e:          20 seconds      Q:
4        Enter normalized frequency increment:       0.01            10
5                       Enter time increment:        0.075
6  ─────────────────────────────────────────────────────────────────────
7  FREQUENCY          0.75     0.76     0.77     0.78     0.79     0.8
8  Power          0.038461 0.041597 0.045126 0.049115 0.053648 0.058823
9  dB             -14.1497 -13.8093 -13.4556 -13.0877 -12.7044 -12.3044
10 Phase (rad)    1.373400 1.365400 1.356735 1.347319 1.337053 1.325817
11 Phase (deg)    78.69006 78.23171 77.73522 77.19573 76.60750 75.96375
12
13 TIME               0    0.075     0.15    0.225      0.3    0.375
14 Amplitude          1 0.887671 0.583393 0.154684 -0.30441 -0.69397
15 exp[-t/te]         1 0.996257 0.992528 0.988813 0.985111 0.981424
16 Power Loss         0 0.204567 0.644764 0.953823 0.877776 0.481597
17 exp[-2t/te]        1 0.992528 0.985111 0.977751 0.970445 0.963194
18 1/e         0.367879 0.367879 0.367879 0.367879 0.367879 0.367879
19
20 The following rows are used for graphing:
ABC-OF-Q.WK1                  UNDO
```

Fig. 5-5. Home screen of the worksheet for experimenting with *Q*. The formula in Cell B8 is displayed at the upper left-hand corner of the screen. Press the PgDn key to see the time-domain screen.

Worksheet Organization

Cell E3: Enter t_e in seconds. (Cell G4 displays the associated value of *Q*.)

Cell E4: Enter the Frequency axis increment in this cell.

Cell E5: Enter the Time axis increment in this cell.

Row 7: This is the Frequency axis. Enter the desired starting value in Cell B7.

Row 8: The power resonance curve is calculated in this row.

Row 9: The power resonance in dB is calculated in this row.

Rows 10 and 11: These calculate the Phase in radians and degrees.

Row 13: This is the Time axis. Enter the starting time in Cell B13.

Row 14: This is the amplitude of a decaying sine wave.

Row 15: This is the envelope of the decaying amplitude.

Row 16: This is the power loss.

Row 17: This is the envelope for energy decay.

Row 18: This is the horizontal line for $1/e$.

The next three figures show graphs connected to this worksheet. Gain and phase in terms of *Q* are shown in the graph named *dB & Phase* , Fig. 5-6. Figure 5-7 shows an example of the graph named *Amplitude vs. t*. The graph named *Power & Phase* is shown in Fig. 5-8.

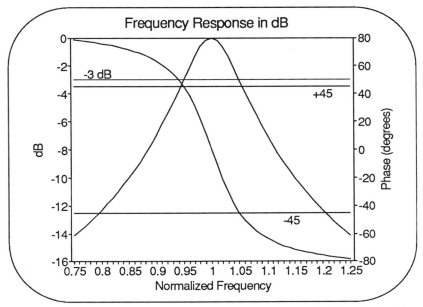

Fig. 5-6. Gain and phase for the Q in Fig. 5-5. Note −3 dB and ±45° lines.

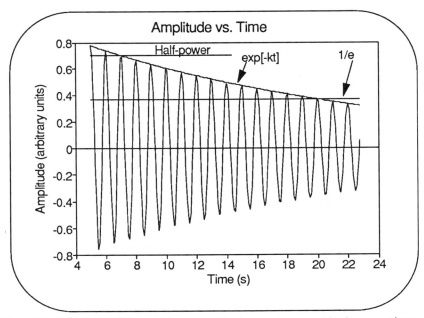

Fig. 5-7. The free decay of the amplitude is shown with the envelope of the exponential damping, the 1/e line, and half-power, for the Q shown in Fig. 5-5.

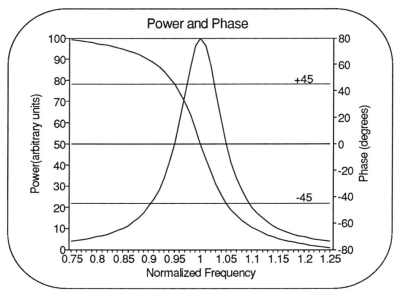

Fig. 5-8. Power and phase for the *Q* in Figs. 5-5, 5-6, and 5-7.

Q-OF-LCR, Worksheet for Series *LCR*

This worksheet lets you input values for a series *L*, *C*, and *R* circuit, and graph the results in serveral forms in the time domain and frequency domain. The Home screen is shown in Fig. 5-9.

Worksheet Organization

Row 4: Enter values of *R*(Ohms) in Cells C4, D4, and E4. *Q* for *R* #1 appears in G4.

Row 5: Enter value of *L* (Henrys) in Cell C5. *Q* for *R* #2 appears in G5.

Row 6: Enter value of *C* (Farads) in Cell C6. *Q* for *R* #3 appears in G6.

Row 9: Undamped resonant frequency is shown in Cells C9 and E9.

Row 11: Frequency axis, normalized to the undamped resonant frequency.

Row 12: Frequency axis in Hz.

Row 13: This row computes $X_L - X_C$ at each frequency.

Rows 14, 15, and 16: These compute the impedance for each value of *R*.

Rows 18, 19, and 20: These rows compute the normalized rms current for each value of *R*.

Row 22: Time axis in units of seconds.

Rows 23, 24, and 25: These compute the waveforms for each value of *R*.

Rows 29 and 30: These show the time constants and *Q*s for each *LCR* combination.

```
B13: 2*@PI*B$12*$C$5-1/(2*@PI*B$12*$C$6)                              READY
```

```
              A        B       C       D       E       F       G
1  Q-OF-LCR.WK1           SERIES LCR CIRCUIT: FREQUENCY AND TIME
2  Chapter 5      Enter data in cells C4, C5, C6, D4, and E4:
3                       #1      #2      #3
4  MORE: PgDn      R:    2.00    5.00   20.00    Q #1:        25
5                  L: 5.00E-03                   Q #2:        10
6                  C: 2.00E-06                   Q #3:       2.5
7
8        Undamped Resonant Frequency:   10000 rad/s  = 1591.549 Hz
9
10 Normalized
11    Frequency      0.8     0.81    0.82    0.83    0.84    0.85
12 Frequency Hz  1273.239 1289.155 1305.070 1320.986 1336.901 1352.817
13 XL - XC         -22.5 -21.2283 -19.9756 -18.7409 -17.5238 -16.3235
14 Z #1         22.58871 21.32240 20.07548 18.84737 17.63757 16.44559
15 Z #2         23.04886 21.80928 20.59186 19.39648 18.22316 17.07212
16 Z #3         30.10398 29.16581 28.26703 27.40846 26.59104 25.81584
17
18 (Irms/Vrms)#1 0.044269 0.046899 0.049812 0.053057 0.056697 0.060806
19 (Irms/Vrms)#2 0.043386 0.045852 0.048562 0.051555 0.054875 0.058575
20 (Irms/Vrms)#3 0.033218 0.034286 0.035376 0.036485 0.037606 0.038735
Q-OF-LCR.WK1   UNDO
```

Fig. 5-9. Home screen of Q-OF-LCR. Press PgDn for the time domain segment.

Rows 30 and 31: These show the resonant frequency for each value of R.

Row 33: This row shows the frequency shift from the undamped frequency, for each *R*. Figures 5-10, 5-11, 5-12, and 5-13 show graphs connected to the worksheet.

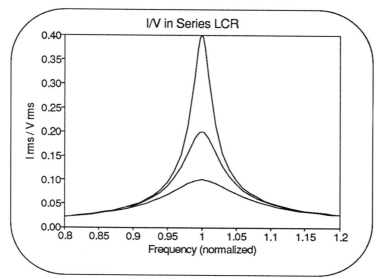

Fig. 5-10. *I*/*V* in the frequency domain for three values of *Q* (see Fig. 5-9).

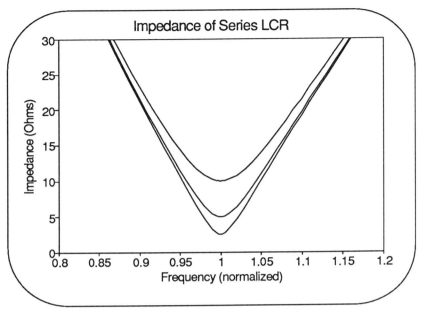

Fig. 5-11. *Z* in the frequency domain for three values of *Q* (see Fig. 5-9).

```
B30: 2*$C$6/$C$5                                                    READY

          A              B       C       D       E       F       G
1   Q-OF-LCR.WK1           SERIES LCR CIRCUIT: FREQUENCY AND TIME
22
23  Time               0  0.00004 0.00008 0.00012 0.00016  0.0002
24  Waveform #1        1 0.913752 0.685761 0.353983 -0.02797 -0.39947
25  Waveform #2        1 0.903013 0.670077 0.342572 -0.02510 -0.37448
26  Waveform #3        1 0.853123 0.603496 0.302730 0.002267 -0.25410
27
28  TIME CONSTANTS AND FREQUENCY SHIFTS DUE TO DAMPING:  (t sub e = 2L/R)
29                     #1              #2              #3
30  2L/R              0.005           0.002           0.0005
31  R/2L              200             500             2000
32  Q                 25              10              2.5
33  Freq., rad/s   9997.999        9987.492        9797.958
34  Freq., Hz      1591.231        1589.558        1559.393
35  Shift, Hz      0.318341        1.990681        32.15582
36
37    Phase of Current:
38  Phase #1       -84.9203 -84.6178 -84.2824 -83.9085 -83.4889 -83.0147
39  Phase #2       -77.4711 -76.7464 -75.9472 -75.0617 -74.0751 -72.9698
40  Phase #3       -48.3664 -46.7066 -44.9650 -43.1386 -41.2245 -39.2205
Q-OF-LCR.WK1           UNDO
```

Fig. 5-12. Second screen of Q-OF-LCR. This calculates the free decay in time for three values of *Q*, and shows center frequency and shift from zero damping.

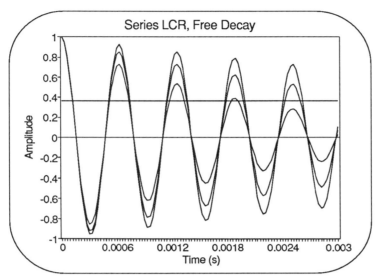

Fig. 5-13. Free decay of damped oscillations for three Q's shown in Fig. 5-9 and 5-12. Upper: $Q = 25$. Middle: $Q = 10$. Lower: $Q = 2.5$.

Q Multipliers

Because Q is decreased by losses in a system it seems natural to try to increase Q by minimizing these losses. Another obvious approach is to add some of the power that was lost by means of active devices. A "Q multiplier" is a device that performs this operation. These are usually electronic circuits that incorporate a positive feedback path and care must be exercised to avoid oscillations.

Answers to Self-Tests

Self-Test 5-1
(a) BW = 175 Hz
(b) Lower –3 dB point = 9912 Hz
(c) Upper –3 dB point = 10,088 Hz

Self-Test 5-2
$$Q = \frac{1000}{1050 - 950}\left[\left(\frac{1}{0.5}\right)^2 - 1\right]^{1/2} = 17.32$$

Self-Test 5-3

$$Q = \frac{1}{2}\left[\frac{(\pi/3 - - \pi/3)}{1.05 - 0.95}\right] = 10.47 \quad \textit{Note}: 60 \text{ degrees} = \pi/3$$

Self-Test 5-4

$$Q = \frac{\pi}{\ln(9.99/9.00)} = 30.1$$

Digging Deeper

"The story of Q," Estill I. Green, *American Scientist* **43**, 584-594 (1955). This paper relates the interesting historical background of Q and its origin in what was called "electric wave filters." Green traces its evolution starting with K. S. Johnson at Western Electric Company's Engineering Department, which in 1925 became Bell Telephone Laboratories. Green's paper, containing references back to 1914, was written before the advent of lasers, superconducting cavities, Q-switches, and the Mössbauer effect, but at the end of the paper Green speculates that ". . . quantum mechanical phenomena . . . seem to offer the best promise for non-aging frequency and time standards . . ."

SSP: The Spreadsheet Signal Processor, S. C. Bloch (Prentice-Hall, Englewood Cliffs, New Jersey, 1992). ISBN 0-13-830506-4. Appendix 1 shows the mathematical interrelationships of Q in terms of the Fourier transform. It's easy but it uses a little calculus. Measurement of Q in the time domain is completely equivalent to measurement in the frequency domain.

Systems with Small Dissipation, V. Braginsky, V. P. Mitrofanov, and V. I. Panov (University of Chicago Press, Chicago, 1985). ISBN 0-226-07073-5. This little book contains a wealth of practical and theoretical information on high-Q systems. In particular it describes ultrahigh-Q microwave cavities made by plating superconducting material on almost-perfect single crystals of sapphire.

. . . when you can measure what you are speaking about and express it in numbers you know something about it; but when you cannot measure it, when you cannot express it in numbers, your knowledge is of a meagre and unsatisfactory kind. . .

Lord Kelvin
in a lecture to the Institution of Civil Engineers
May 3, 1883

Chapter 6

Low-Pass and High-Pass Filters

Like a tea bag, a coffee filter, or a swimming-pool filter, an electronic filter can be used to select what passes through and what is taken out. The decibel is extremely useful in describing the characteristics of filters. In this chapter we will discuss:

- Four specialized definitions of the decibel
- The concept of Gain
- The frequency and time characteristics of simple low-pass and high-pass filters.

The filters that we will look at are "passive" filters in the sense that no additional energy is supplied by the filter; all energy comes from the input signal. Worksheets are provided so you can become familiar with the characteristics of these filters by experimenting with them on your personal computer. The worksheets have graphs set up and ready to run; the graphs are "live", connected to their respective worksheets so you can instantly view and print the results of your experiments.

dB, dBW, dBm

The fundamental definition of the decibel is in terms of power,

$$dB = 10\log(P_o/P_r)$$

where P_o is the measured power output and P_r is the reference power. It is often convenient to use one Watt as the reference; this is so common that the decibel referred to one Watt is designated as dBW,

$$dBW = 10\log(P_o/1W).$$

Another common power reference is one milliwatt. The decibel designation for a one milliwatt reference is dBm,

$$dBm = 10\log(P_o/10^{-3}W).$$

You will always obtain correct results if you measure power levels to obtain decibels. Remember that the power is conventionally measured in terms of its rms value, not its instantaneous value, because rms values make measurements independent of the shapes of the waveforms. On the other hand, voltmeters and oscilloscopes are more common instruments than power meters so it is necessary to relate decibel measurements to voltage. If a voltage V exists across a resistance R then the conversion from electrical power to thermal power in the resistance is $P = V^2/R$, where P is measured in Watts, V in Volts, and R in Ohms. See Appendix 1, Figs. A1-1 and A1-2, for graphs of dB vs. voltage and power ratios.

Since V varies in time, it is conventional to use rms values of V; therefore P will be its rms value. This is also true if R is replaced by an impedance of Z Ohms. In terms of voltage and impedance we can write the decibel as,

$$dB = 10 \log \left(\frac{V_o^2/Z_o}{V_r^2/Z_r} \right).$$

So, if the output Z is identical to the reference Z they cancel and we have

$$dB = 10 \log(V_o/V_r)^2.$$

According to the rule of logarithms,

$$\log(x^n) = n \log(x)$$

so the expression above may be written,

$$dB = 20 \log(V_o/V_r).$$

V_o is a measured output voltage and V_r is the reference or input voltage. For the above expression to be true both voltages must be measured across the same value of impedance! Remember, the way we obtained this expression from power is that we cancelled the impedances in the numerator and denominator and they must have the same value to cancel! A common mistake is to measure voltages across different input and output impedances and then put these voltages in the expression for dB. Don't do it.

Decibel Measurement Across Different Impedances

Power is the product of voltage times current,

$$P = VI.$$

Ohm's Law relates voltage, current, and impedance, $V = IZ$, so we can express power as,

$$P = \frac{V^2}{Z} = I^2 Z$$

In order to express the power ratio of the output P_o to the reference P_r for decibel measurements we can use the equation above and write,

$$P_o/P_r = (V_o/V_r)^2 (Z_r/Z_o).$$

In terms of current,

$$P_o/P_r = (I_o/I_r)^2 (Z_o/Z_r).$$

When we take the log of the above expression for the voltage ratio the product of the two parentheses turns into a sum of logs,

$$dB = 20 \log(V_o/V_r) + 10 \log(Z_r/Z_o)$$

When we take the log of the expression for the current ratio we obtain,

$$dB = 20 \log(I_o/I_r) + 10 \log(Z_o/Z_r)$$

The term involving the log of the impedances goes to zero *only* in the special case when $Z_o = Z_r$. See Fig. A1-2 in Appendix 1 for a graph of dBm vs. voltage for different impedances.

dBV

The reference voltage can be whatever you choose. For this reason some people say you use dB's when you don't know what you're measuring; this is not correct because a measurement in dB's is a relative measurement. One volt and one millivolt are common references; dBV usually means $V_r = 1$ Volt. See Appendix 1 for some other common, and uncommon, references. For example, suppose at a frequency of 1 kHz an amplifier provides an output of 15 V for an input of 1 V, with both voltages measured across identical impedances. What is the output in dBV relative to the input?

$$dBV = 20 \log(15/1) = 20 \log(15) = 23.5.$$

This is called the gain at 1 kHz, in dBV.

Gain in dB

Power gain is defined as.

$$Gain = Power\ Out/Power\ In$$

So,

$$\text{Power Out} = \text{Gain} \times \text{Power In.}$$

Taking logarithms of both sides of the previous expression and multiplying by 10 to obtain Gain in dB, we have,

$$\text{Gain (dB)} = 10 \log(\text{Power Out/Power In}).$$

Now, since power is proportional to $(\text{voltage})^2$,

$$\text{Gain (dB)} = 10 \log(\text{voltage out/voltage in})^2$$

and therefore, since $\log(x^n) = n \log(x)$,

$$10 \log(x^2) = 20 \log(x)$$

we have,

$$\text{Gain (dB)} = 20 \log(\text{voltage out/voltage in}).$$

This is a very useful expression for calculating Gain based on voltage measurements. Remember, in the above expressions the voltage out and the voltage in must be measured across identical impedances; otherwise we must add the correction term involving the log of the impedance ratio. Gain can be written in another useful form using the rule of logarithms that $\log(A/B) = \log A - \log B$,

$$\text{Gain} = \text{Output} - \text{Input} \quad \textit{(all in dBs)}$$

or, finally,

$$\text{Input} + \text{Gain} = \text{Output.} \quad \textit{(all in dBs)}$$

Examples of use of Gain expressed in dB's:

If the input is −10 dB and the gain is 60 dB, what is the output?
Output = Input + Gain = −10 dB + 60 dB = 50 dB.

If the input is 20 dB and the output is 80 dB, what is the gain?
Gain = Output − Input = 80 dB − 20 dB = 60 dB.

Gain and Phase of Simple Low-Pass Filters

A low-pass filter is an electronic, mechanical, or acoustic system that passes low frequencies much better than high frequencies. The so-called *cut-off frequency* of a low-pass filter is arbitrarily taken to be the frequency corresponding to the −3 dB point. The gain of a passive filter is always expressed as negative dBs because the output is always less than the input. The gain of an active filter (one that adds energy) may have positive gain over a wide range of frequencies, while suppressing other frequencies. Here we will only discuss passive filters. At frequencies well below the cut-off frequency there is very little attenuation in a low-pass filter. At the −3 dB point the output has fallen to half the input power, and above the cut-off frequency the output drops very rapidly with increasing frequency. This definition is shown graphically in Fig. 6-1 which presents the gain and phase shift measured with sine waves over a wide frequency range.

A simple low-pass electronic filter can be constructed with a resistor and capacitor, as shown in Appendix 5, Fig. A5-4. So we won't interrupt our stream of consciousness with details, all of the analysis of this filter is securely tucked away in Appendix 5. Only algebra is used in Appendix 5; you can brush up on your algebra in Appendix 4.

The filter shifts the phase of the output backwards relative to the input, and this phase shift depends on frequency as shown. At low frequencies there is practically no phase shift; at the −3 dB point the phase shift is −45 degrees, and at very high frequencies the phase shift approaches −90 degrees asymptotically. Fig. 6-2 shows the sine wave response in the time domain.

Measuring the response of a filter with sine waves gives one picture of filter characteristics, in the frequency domain. In order to complete the filter picture it is also necessary to look at the response in the time domain. Other convenient test signals in the time domain are rectangular pulses and square waves; from a Fourier perspective both test signals contain many frequencies.

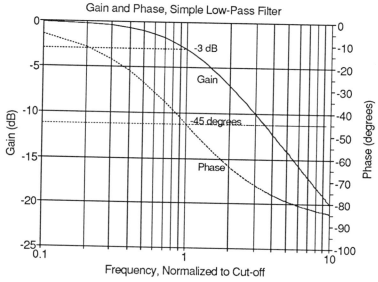

Fig. 6-1. Gain at the cut-off frequency is −3 dB and the phase shift is −45 degrees for a simple filter. Also see Fig. A5-4 in Appendix 5.

Because the output of a low-pass filter is deficient in high frequencies (and phase shift depends on frequency, too) the pulse or square wave will be distorted by the filter in the time domain. In Fig. 6-3 we see the output of the filter plotted as a function of the input, with time as a parameter. Figs. 6-2 and 6-3 are just different representations of the same information.

The time for the voltage to fall to $1/e$ of its maximum value is called the time constant or $1/e$ folding time for voltage; we designate this as t_e. The time for the voltage to fall to $1/2$ of its maximum value is called the half-life for voltage; we designate this as t_h. The time constant and half-life are measured in seconds. There are simple relationships between these measures of decay times, cut-off frequencies, and the parameters of the systems. These are summarized for the elementary filters discussed here, in Table 6-1.

Table 6-1. Frequency and Time Parameters for Simple Low- and High-Pass Filters		
Cut-off Frequency (Hz) *(–3 dB), f_c*	*Time Constant (s)* *(1/e), t_e*	*Half-Life (s)* *(1/2), t_h*
$1/(2\pi RC)$	RC	$0.693147 t_e$
	$1.442695 t_h$	

Note 1: For resistor-inductor filter, substitute L/R for RC.
Note 2: $\ln(2) = 0.693147...$
Note 3: R in Ohms, C in Farads, L in Henrys

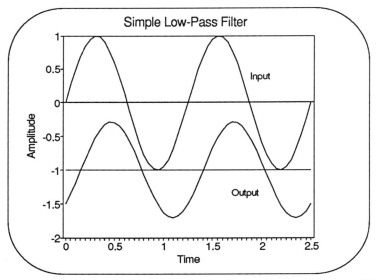

Fig. 6-2. Time domain response of the low-pass filter at the cut-off frequency. Note amplitude decrease (.707 of input) and phase shift (–45°) of the output.

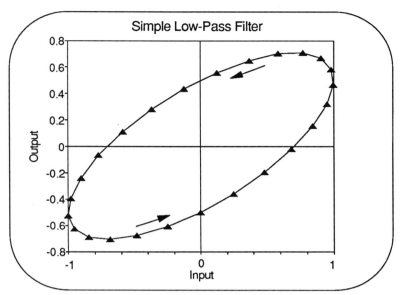

Fig. 6-3. Output vs. input at the cut-off frequency. This ellipse has the same information plotted in Fig. 6-2. Time increment between markers is constant.

The exponential rise and decay of the low-pass filter can be visualized in a simple way be using a rectangular pulse as the input signal. This is shown in Fig. 6-4.

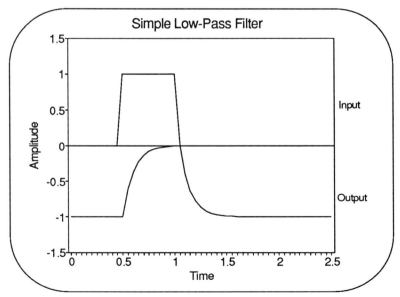

Fig. 6-4. Typical time domain response of a low-pass filter for a rectangular pulse input. The output has been displaced by −1 for clarity.

A low-pass filter cleans up a signal by reducing high-frequency noise (variance reduction) but it also produces some undesirable effects. In Fig. 6-5 the solid line is the input, the dotted line is the output. The lack of highs in the low-pass output results in loss of details; sharp corners become rounded (technically known as *round-off*) and rapid changes can not be preserved. The output is approximately the integral (in the calculus sense) of the input, so low-pass filters are often called *integrators*. The lack of lows in the high-pass output (shown in the next section) results in the inability to sustain the amplitude for a long time (this is technically known as *droop*) but rapid changes are preserved. A pulse or square wave test signal gives a convenient quick look at the frequency response of a system, in terms of its time response.

LOW-PASS, Low-Pass Filter Worksheet

You can experiment with low-pass filters by changing their parameters and observing the results. Retrieve the worksheet LOW-PASS in your spreadsheet, and you should see the Home screen shown in Fig. 6-6.

Several graphs are available in this worksheet. These graphs are "named" and you can view them in the usual way using your spreadsheet commands. For example, if your spreadsheet is compatible with *Lotus 1-2-3* commands, you can view the graph named *Gain(dB)* by using the command /Graph Name Use and choosing the graph name by placing the cursor on it and pressing Enter⏎ .

You should see something like the graph shown in Fig. 6-7. It may look different, depending on your particular spreadsheet.

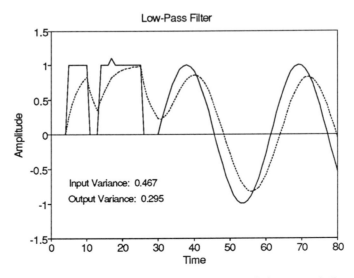

Fig. 6-5. Low-pass filter decreases amplitude and time resolution for pulses, and produces phase and amplitude distortion of sine wave. Input: Solid line.

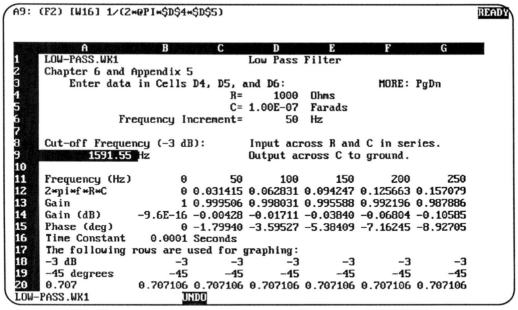

```
A9: (F2) [W16] 1/(2*@PI*$D$4*$D$5)                                    READY

        A           B        C        D        E        F        G
1   LOW-PASS.WK1                    Low Pass Filter
2   Chapter 6 and Appendix 5
3       Enter data in Cells D4, D5, and D6:              MORE: PgDn
4                               R=     1000   Ohms
5                               C= 1.00E-07   Farads
6           Frequency Increment=        50   Hz
7
8   Cut-off Frequency (-3 dB):         Input across R and C in series.
9        1591.55  Hz                   Output across C to ground.
10
11  Frequency (Hz)        0       50      100      150      200      250
12  2*pi*f*R*C            0 0.031415 0.062831 0.094247 0.125663 0.157079
13  Gain                 1 0.999506 0.998031 0.995588 0.992196 0.987886
14  Gain (dB)       -9.6E-16 -0.00428 -0.01711 -0.03840 -0.06804 -0.10585
15  Phase (deg)          0 -1.79940 -3.59527 -5.38409 -7.16245 -8.92705
16  Time Constant   0.0001 Seconds
17  The following rows are used for graphing:
18  -3 dB                -3       -3       -3       -3       -3       -3
19  -45 degrees         -45      -45      -45      -45      -45      -45
20  0.707          0.707106 0.707106 0.707106 0.707106 0.707106 0.707106
LOW-PASS.WK1          UNDO
```

Fig. 6-6. Home screen for simple low-pass filters. See Fig. A5-4 for circuit.

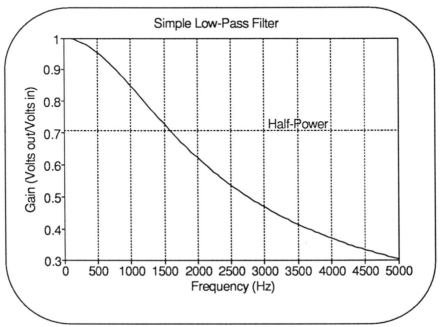

Fig. 6-7. Worksheet graph for Gain on linear scales. Graph name: *Gain*.

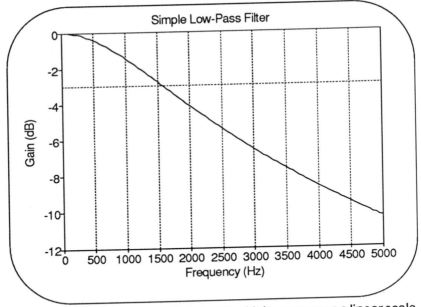

Fig. 6-8. Worksheet graph for Gain in dB, with frequency on a linear scale. Graph name: *Gain [dB]*.

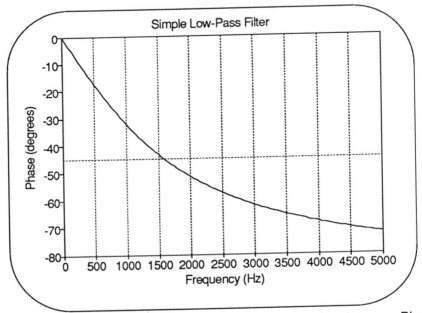

Fig. 6-9. Worksheet graph for phase on linear scales. Graph name: *Phase*.

Gain and Phase of Simple High-Pass Filters

A high-pass filter passes high frequencies much better than low frequencies. Below the cut-off frequency little power gets through the filter; above the cut-off frequency almost all of the input flows through to the output. A typical frequency response of a simple high-pass filter is shown in Fig. 6-10.

The phase shift of this high-pass filter is opposite to that of the low-pass filter. At low frequencies the phase is asymptotic to +90 degrees, at the −3 dB point the phase is +45 degrees, and the phase asymptotically (that word again!) approaches 0 degrees at very high frequencies. Again, to complete the picture we will look at the responses to sines and pulses in the time domain (see Fig. 6-11). The output is distorted because of the phase shifts and the output is deficient in low frequencies compared with the input. Figure 6-12 shows the output as a function of the input, for a sine wave at the cut-off frequency.

The filter response to a rectangular pulse input is shown in Fig. 6-13. The retention of the high-frequency components of the pulse results in fast rise times of the output, but the droop is severe because of the deficit in the low frequencies. The lows are needed to sustain the flat-top of the pulse. The output is an approximate derivative (in the calculus sense) of the input, and this is why high-pass filters are sometimes called *differentiators*.

The polar plots in Figs. 6-3 and 6-12 reveal how the low-pass and high-pass filters accomplish their respective missions. The low-pass filter retards the phase and the high-pass filter advances the phase.

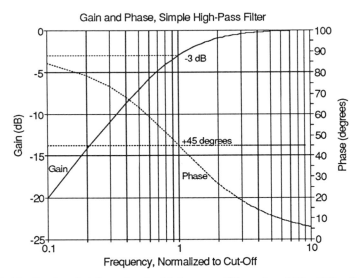

Fig. 6-10. Bode plot for simple high-pass filter. $f_c = 1/(2\pi RC)$. At the cut-off frequency the gain is −3 dB and the phase shift is +45 degrees.

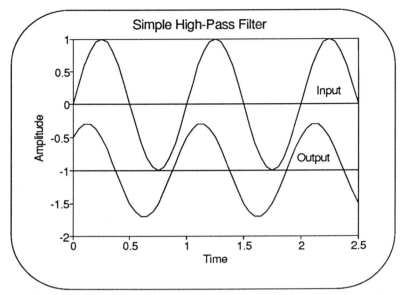

Fig. 6-11. Time domain response of the high-pass filter at the cut-off frequency. Observe the output amplitude decrease (0.707 of input) and the phase shift (+45 degrees relative to the input). Compare with Fig. 6-2.

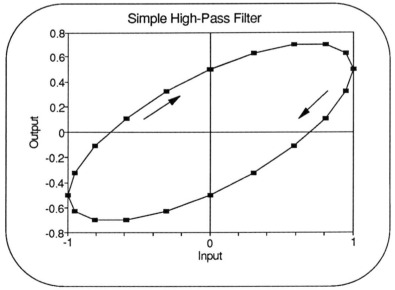

Fig. 6-12. Steady-state output vs. input at the cut-off frequency. This ellipse has the same information plotted in Fig. 6-11. The time increment between markers is constant. Compare with Fig. 6-3; the only change is the phase shift.

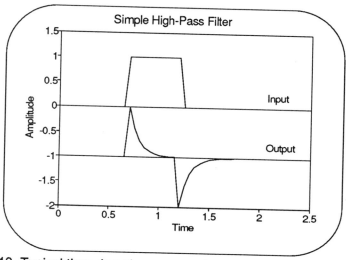

Fig. 6-13. Typical time-domain response for a rectangular pulse input. This is sometimes called an edge detector in image processing.

HI-PASS, High-Pass Filter Worksheet

You can experiment with high-pass filters with this worksheet. The Home screen is shown in Fig. 6-14. Fig. A5-5 (Appendix 5) has the circuit diagram and its analysis. Figs. 6-15, 6-16, and 6-17 show graphs connected to this worksheet.

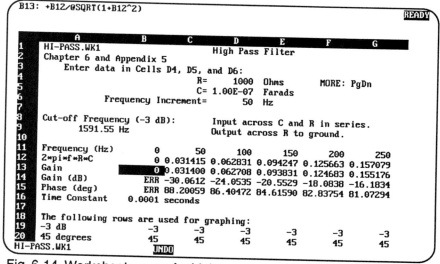

Fig. 6-14. Worksheet screen for high-pass filter. The highlighted cell B13 shows the formula for the Gain at each frequency.

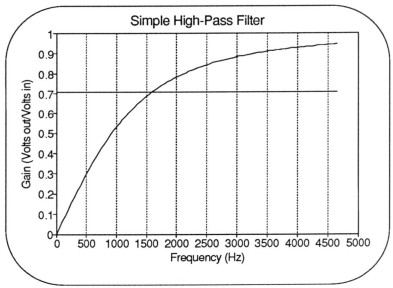

Fig. 6-15. Worksheet graph for Gain on linear scales. The horizontal line at a voltage ratio of 0.707 corresponds to the half-power (−3 dB) line.Graph name: *Gain*.

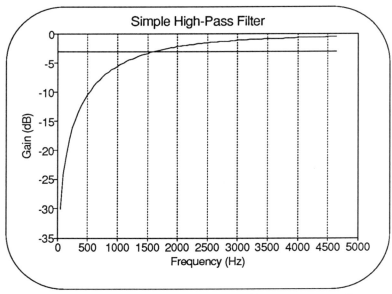

Fig. 6-16. Worksheet graph for Gain in dB and frequency on a linear scale. The horizontal line is −3 dB. Graph name: *Gain [dB]*.

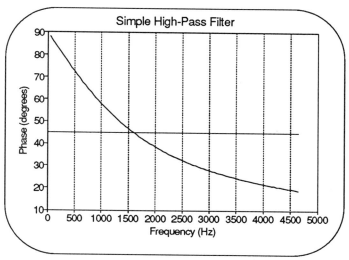

Fig. 6-17. Worksheet graph for phase on linear scales. The horizontal line is +45 degrees. Graph name: *Phase*.

Applications of Low-Pass and High-Pass Filters

A simple filter cannot remove noise in the frequency range where the signal exists, but fortunately in many practical cases the signal spectrum is much more restricted than the noise spectrum. For example, if the signal is confined mainly to low frequencies then a low-pass filter can be used to remove noise at higher frequencies. This can markedly improve the signal/noise ratio. The price you pay is familiar: Amplitude and phase distortion. This can be seen in Fig. 6-18 for pulses (broad-band signals) and a sine wave (narrow-band signal).

The simple filters we have discussed can be made more sophisticated by applying them more than once. The output of one filter can be sent to another filter, and on and on. This can steepen the roll-off of the Gain with frequency, but the price paid for this is to increase the rate of change of phase near cut-off. This can have undesirable consequences in feedback control systems. The use of two filters is shown in Fig. 6-19.

A high-pass filter is useful when the noise and interference spectrum lies mostly below the signal spectrum. This often occurs in semiconductors where the so-called "1/f" noise is a problem (see Chapter 11). This type of noise increases as frequency decreases, so it is amenable to reduction by a high-pass filter.

Another common problem is low-frequency coherent (not random) interference, such as power-line hum. Of course, a band-stop filter could be used to reject narrow-band interference like 60 Hz hum, but a high-pass filter will reduce *all* of the low-frequency spectrum.

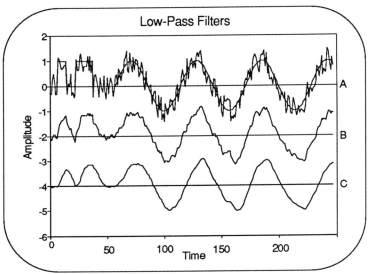

Fig. 6-18. Low-pass filter enhances signals by reducing high-frequency noise, at the expense of decreased horizontal and vertical resolution for pulses, and amplitude and phase errors for a sinewave. A: Input signal and noise. The noiseless signal is superimposed on the noisy signal. B: Output of the first low-pass filter. C: Output of the second low-pass filter.

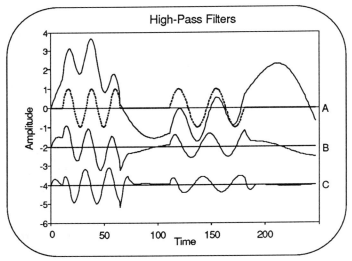

Fig. 6-19. A high-pass filter recovers two pulsed sines from low-frequency coherent interference. A: Signal and interference. Desired signal is shown as dotted line. B: Output of the first high-pass filter. C: Output of the second high-pass filter. Observe that the higher-frequency sine pulse is enhanced more than the lower-frequency sine pulse.

DIGIFILT, Digital Filter Worksheet

This worksheet can be used in two ways. First, you can use it to experiment with digital low-pass and high-pass digital filters with the internally generated signal and noise. Second, you can use these filters with real-world data by loading the data in the worksheet via keyboard, reading a disk file, or by means of a data acquisition board. With a data acquisition board, or using the dynamic data exchange (DDE) feature of Windows, this worksheet can make your computer into a versatile signal processor.

The subject of digital signal processing is vast and still rapidly expanding, and too much to be covered even in several books. Digital filters can perform remarkable feats that are not possible with analog circuits. This worksheet will introduce you to simple processors of the type known as finite impulse response (FIR) filters and infinite impulse response (IIR) filters. The ones in this worksheet are easy to understand from the viewpoint of weighted moving averages, integrators, and differentiators. You have already seen in LOWPASS and HIGHPASS that low-pass filters act like integrators in their smoothing action, and high-pass filters act a bit like differentiators in their detection of changes and disregard of relatively constant signals.

The other general type of digital filter is called infinite impulse response or IIR. These are usually a bit more sophisticated and more difficult to understand than FIR filters. The worksheet DIGIFILT has a simple IIR low-pass filter included so you can experiment with it. This one is so small that it can virtually be hidden in a spreadsheet.

Because digital filters operate in discrete time and frequency the sampling rate of a signal has a profound influence. (See Time Increment in Cell E5 in Fig. 6-20; the smaller the time increment, the more samples per unit time.) You will discover some of these fundamentals as you vary the time increment and sine wave test frequency. A basic concept is that you must sample a signal *at least* twice the rate of the highest frequency that you want to know with confidence. For compact discs the highest frequency (called the Nyquist frequency) is 22.050 kHz so you must collect at least 44,100 samples per second. It's better if the sampling rate is higher. You often see advertisements for compact disc players that say "4 times oversampling" or "8 times oversampling."

What happens if you undersample? You can see one effect, called *aliasing*, in movies about the Old West in which the stagecoach wheels appear to rotate forward or backward slowly. Movies are discrete time, sampled data systems, usually with a frame rate of 24 frames/sec. Aliasing occurs because a frequency above the Nyquist frequency masquerades as a lower frequency. (In movies the Nyquist frequency is half the frame rate.) To reduce this problem in sound and video signals *anti-aliasing filters* are utilized. These are essentially low-pass filters with linear phase characteristics. They reduce power at frequencies above the Nyquist frequency.

You can expect a revolution in movie technology, and image technology in general, brought about by the digital methods of high definition television (HDTV). This is similar to the changeover from analog audio to digital audio tapes and compact discs. In the USA, Canada, and Japan, before HDTV the NTSC video standard was 30 frames/sec with 525 lines/frame. In Europe the standard (PAL) was 25 frames/sec with 625 lines/frame. HDTV has the potential of bringing photographic quality to television and replacing film technology in movies. Digital image processing is now providing results that would have been impossible with matte and mask artistry. George Lucas (Industrial Light & Magic) and Steven Spielberg have used these methods to lower production costs while improving the final work. Perhaps digital holograms are next.

Worksheet Organization

Enter the desired parameters in Cells E4, E5, and E6. Cell E4 sets the time constant for the low-pass and high-pass FIR filters. Cell E5 sets the increment for the Time Axis, Row 8. Cell E6 sets the frequency of the sinewave part of the built-in test waveform. Figure 6-20 shows the Home screen.

Row 8 is the Time Axis.

Row 9 generates a noiseless internal test signal. The diskette comes with a rectangular pulse and a sine wave.

Row 10 enables you to add random noise and other signals to the test signal.

Row 11 composes the final internal test signal, made of signal and noise.

Cells B14 through H14 show the time-reversed finite impulse response, known as h(t)REV. Cell I14 contains the formula @SUM($B14...$H14). This is the normalization factor for the filters.

Row 16, starting with Cell H16, is the output of the FIR low-pass filter. See Fig. 6-20 for the formula for the first point of the output of this filter.

Row 17, starting with Cell H17, is the output of the FIR high-pass filter.

Cell C19 lets you enter the k factor for the IIR low-pass filter. As this factor increases the cut-off frequency decreases. In this filter the k factor must be greater than 1.

All of these filters are of the so-called single-pole variety. It is easy to use the output of one of these filters as the input another of the same, or different, type. By this means you can construct rather sophisticated filters of the multi-pole type. For example, Sending the output of a low-pass filter to another low-pass filter will steepen the roll-off after the cut-off frequency, but the rate of change of phase with frequency will also increase.

Figure 6-21 shows an input and output of the FIR filter for wide-band (pulse) and narrow-band (sine) signals. Notice the "learning time" for the sine wave; the filter requires 7 time units to achieve the proper amplitude and phase.

H16: (+$B14*B11+$C14*C11+$D14*D11+$E14*E11+$F14*F11+$G14*G11+$H14*H11)/$I$1 **EDIT**
(+$B14*B11+$C14*C11+$D14*D11+$E14*E11+$F14*F11+$G14*G11+$H14*H11)/$I$13

	A	B	C	D	E	F	G	H	I
1	DIGIFILT.WK1			Low-Pass and High-Pass Digital Filters					
2	Chapter 6								
3	Enter values in Cells E4, E5, and E6:								
4				Time Constant=	2	Seconds		MORE: PgDn	
5				Time Increment=	1	Seconds			
6				Sine Frequency=	0.05	Hz			
7									
8	TIME	-6	-5	-4	-3	-2	-1	0	1
9	Test Signal	0	0	0	0	0	0	0	0
10	Noise	0	0	0	0	0	0	0	0
11	S + N	0	0	0	0	0	0	0	0
12									
13	FIR Filters:					@SUM($B$14..$H$14):		2.4647	
14	h(t)REV	0.0497	0.0820	0.1353	0.2231	0.3678	0.6065	1	
15	OUTPUTS:								
16	Low-pass	0	0	0	0	0	0	0	0
17	High-pass	0	0	0	0	0	0	0	0
18									
19	IIR Filter. Set k=		5						
20	Low-pass	0	0	0	0	0	0	0	0

DIGIFILT.WK1

Fig. 6-20. Home screen of DIGIFILT worksheet. Cell H16 shows the formula for the output of the low-pass FIR filter at time $t = 0$. For the IIR filter, enter the k-value in Cell C19; this value must be greater than 1.

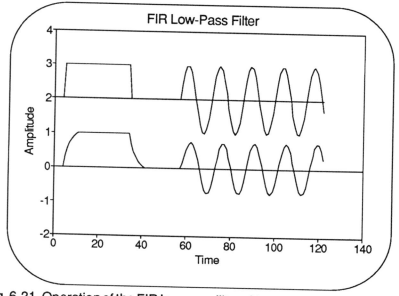

Fig. 6-21. Operation of the FIR low-pass filter. Observe the round-off of the pulse output, indicative of high-frequency deficit. Also note the amplitude decrease and phase shift of the sine wave. Graph name: *FIR Low-pass*.

How the filters work: To get an idea of the operation of the FIR low-pass filter examine the formula shown in Cell H16 in Fig. 6-20 above. Cells B14..H14 contain a time-reversed exponential decay called $h(t)REV$. The time constant for this is set by Cell E4. The sum of the values in Cells B14..H14 is contained in Cell I14. The formula for Cell H16 indicates that the value in this cell is the input signal (Cell H11) modified by the exponentially weighted average of the the previous six input values, all normalized to the sum of the weights.

Now you can see how this behaves as an integrator; it smooths a finite number of values by a moving average, reducing rapid variations. It "forgets" rapidly. Better frequency response characteristics can be obtained by using more points in $h(t)REV$, at the expense of more computation. Observe the use of the absolute cell address (like $B14) in the formula. The instructs the worksheet to drag $h(t)REV$ through the input signal. The worksheet has a built-in test signal that you can modify at will. At present it consists of a rectangular pulse and a sine wave.

With the worksheet running on your computer you can browse through the other Cells. Look at the formula in Cell I17, the first output point of the high-pass filter. In this case Cell H17 is not the first output point because a high-pass filter looks at the difference between points. Here the first difference is between Cell I17 and the sum of the previous six cells, exponentially weighted and normalized. The ultimate high-pass filter possible on this worksheet is simply the difference between two adjacent cells, divided by the time increment. Now you can see why a high-pass filter acts like a differentiator or an operator that takes the first derivative.

The IIR low-pass filter is more sophisticated but it actually appears to be simpler. The IIR filter uses a *feedback* path in the following way: It computes a weighted average between the contemporary input signal and the previous output signal by the formula $[(1/k) \times \text{current input}] + (1 - 1/k) \times \text{previous output}$. The result is that the contemporary output of the filter depends on the contemporary input and all of the previous input, potentially of infinite duration. That's why this is called an infinite impulse response filter. When $k = 1$ the output is equal to the input, and when $k > 1$ filtering action occurs. For example, when $k = 2$ the input and the previous output are weighted equally. For the value $k = 5$ shown in the worksheet, the input is weighted 0.2 and the previous output is weighted 0.8 so the cut-off frequency decreases as k increases. Try a value of $k < 1$ and you will see the filter break into oscillation, a common occurrence in feedback systems.

Another useful way to view the output of the filter is by plotting the output vs. the input with time as a parameter. We have seen this before for the steady-state case but now you can see the beginning of the output as the filter gradually acquires its final amplitude and phase. Because this FIR filter has seven points it takes seven time units to "learn" the proper amplitude and phase. Fig. 6-22 provides some additional insight into the operation of the filter. Observe the start of the signal and see how the phase takes off counter-clockwise, indicating a phase lag. This is characteristic of a low-pass filter. Figs. 6-23 and 6-24 show the FIR high-pass.

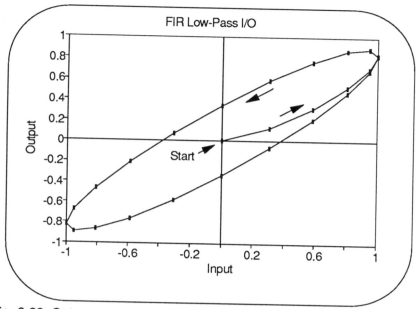

Fig. 6-22. Output vs. input for the FIR low-pass filter. The markers are evenly spaced in time. Graph name: *FIR Low-pass I/O*.

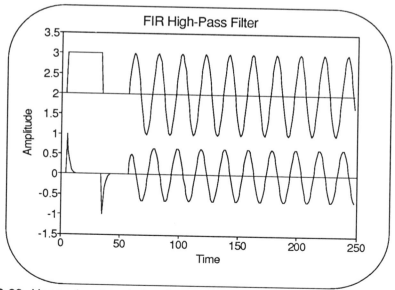

Fig. 6-23. Upper: Input test signal. Lower: output for the FIR high-pass filter. Observe the differentiation of the pulse, and the amplitude decrease and phase lead of the sine wave. Graph name: *FIR High-pass*.

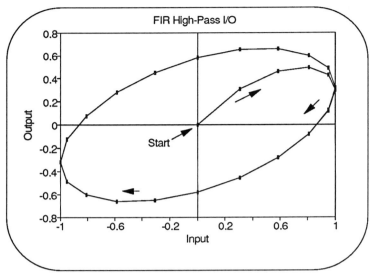

Fig. 6-24. Output vs. input for the FIR high-pass filter. After the start of the signal the curve begins its clockwise motion, indicating a phase lead. The Input and Output scales are different! Compare with Fig. 6-22. Name: *FIR High-pass I/O*.

The graph named *IIR Low-Pass* (Fig. 6-25) is not much different from that of the FIR low-pass filter in Fig. 6-21, except that the IIR filter output is much smoother.

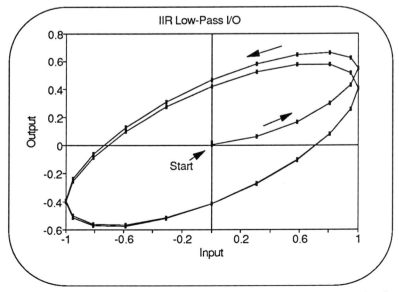

Fig. 6-25. Counter-clockwise rotation indicates a phase lag for the low-pass. Note gradual transition to the steady-state amplitude and phase, typical of IIR.

DIGISWEP, Frequency Sweep Worksheet

This worksheet is similar to DIGIFILT except that the test signal is swept in frequency. Retrieve this worksheet and look at the input and output waveforms for the low-pass and high-pass filters as the frequency changes. You can vary the starting frequency and the sweep rate by entering the desired values as usual. When you view the I/O plots you may be pleasantly surprised, especially if your computer is slow. Even on a 66 MHz computer you can see the semi-animated plot as the amplitudes and phases shift. Turn off the "turbo" switch (if you have one) to view these graphs more slowly.

The frequency sweep is accomplished by modifying the formula for a constant-frequency sine wave test signal, $A \sin(\omega_o t + \theta)$. Here, A is the amplitude, ω_o is the constant frequency, and θ is the phase angle. To modify this to sweep in frequency, starting at ω_o with a sweep rate R radians/sec^2, use the formula

$$A \sin(\omega_o t + 0.5Rt^2).$$

This will produce a linear FM sweep, that is, the frequency will increase directly proportional to time. When you use a test signal like this remember that the higher frequencies are sampled less often than the lower frequencies, so you have to be careful about aliasing. You may need to modify the sampling rate and also restrict the highest frequency of the sweep to prevent aliasing.

Digging Deeper

PC-DSP, O. Alkin (Prentice-Hall, Englewood Cliffs, New Jersey, 1990). ISBN 0-13-655200-5. This book, with its included disks, gives a brief introduction to many of the fundamental digital signal processing operations. Included are convolution, auto- and cross-correlation, Fast Fourier Transforms, and design and analysis of FIR and IIR digital filters.

Introduction to Digital Signal Processing, a Computer Laboratory Textbook, M. J. T. Smith and R. M. Mersereau (John Wiley & Sons, New York, 1992). ISBN 0-471-51693-7. This computer lab book and its disks provide many of the basic concepts of digital signal processing. The chapters are keyed to standard textbooks for more complete discussions and derivations.

The Society of Motion Picture and Television Engineers (*SMPTE*) is a good source for the latest developments in digital audio and image processing applications in the entertainment industry. As Dolby and THX have revolutionized movie audio, computer image processing is revolutionizing what you see. Look for the *SMPTE Journal* at the library. Also, check Digging Deeper at the end of Chapter 9.

Chapter 7

Band-Pass and Band-Stop Filters

In this chapter we will examine the frequency and time characteristics of simple passive band-pass and band-stop filters. By "passive" we mean that no extra energy is added to the filter. The only energy input is that of the input signal.

In Chapter 4 we examined the general concept of the Q in some detail because it provides insight into the workings of these filters, from a different point of view.

A band-pass filter can select and pass a relatively narrow band of frequencies, reducing the energy in signals above and below its resonant frequency. In Chapter 4 we discussed acoustic resonances of enclosed volumes; you can think of a room as a collection of 3-dimensional band-pass filters. Electronic band-pass filters are useful in several applications such as:

- Tuning a radio or TV to select a narrow band or a single channel

- Reducing noise and harmonic content on a sine wave from a signal generator. (See Chapter 9)

- Obtaining an estimate of a spectrum, using a bank of band-pass filters or by slowly tuning through a range of frequencies (see Chapter 8)

A band-stop (or notch) filter has the property of passing all frequencies *except* those in a relatively narrow band. A band-stop filter is useful for:

- Rejecting power-line hum

- Rejecting an interfering signal

- Rejecting the fundamental frequency of a signal so the harmonic content can be measured, as in measurement of Total Harmonic Distortion (THD)

By using a combination of band-pass and band-stop filters you can enhance a desired signal, but (as will be seen) you have to be careful about the inevitable phase and amplitude distortion that they introduce.

Worksheets are provided on the diskette so you can experiment with the properties of band-pass and band-stop filters.

Gain and Phase of Simple Band-Pass Filters

A band-pass filter can be thought of as a combination of low-pass and high-pass filters arranged to pass a band of frequencies very well, but to reject frequencies outside of that band. The frequency interval between the –3 dB points on the frequency response curve is called the bandwidth of the filter. The bandwidth is also known as the Full Width at Half Power (FWHP), or the Full Width at Half Maximum (FWHM).

The gain and phase characteristics of a simple *LCR* circuit, used as a band-pass filter, are shown in Fig. 7-1 in dB and degrees vs. log frequency (a Bode plot). Appendix 5 for a typical circuit diagram of this filter and its analysis.

A band-pass filter has a maximum response at its resonant frequency but it has significant response at frequencies away from resonance. The sharpness of the resonance is determined by the *Q* of the filter (see Chapter 4) which is determined by the ratio of the energy stored to the energy lost, per half cycle. In other words, a low-loss filter will have a high *Q* and will respond to a relatively narrow band of frequencies.

Bandwidth is usually specified as Hz, kHz, or MHz FWHM, except for hearing aids which have their own standards (see page 128).

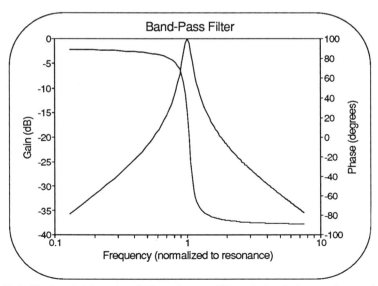

Fig. 7-1. Bode plot for simple band-pass filter. Actual shape depends on *Q*.

The *skirt selectivity* is an important parameter of more sophisticated filters, and it is an indication of how fast the sides or skirts of the response curve fall as the frequency moves out from the –3 dB points. The skirt selectivity is often specified in terms of the –30 dB and –60 dB points.

To complete the picture of the band-pass filter, Fig. 7-2 shows a typical response in the time domain.

BANDPASS, Band-Pass Filter Worksheet

Retrieve BANDPASS in the usual way and you should see the Home screen shown in Fig. 7-3. The circuit diagram is shown in Fig. A5-6, Appendix 5.

Worksheet Organization

Cells D4, D5, D6, and D7: Enter information on resistance, inductance, capacitance, and frequency increment.

Cell A9: The resonant frequency (Hz) is calculated and displayed in this cell.

Row 15: This row is the frequency axis. Enter the desired starting frequency in Cell B15.

Row 16: This row computes $\omega L/R$ at each frequency.

Row 17: This row computes $1-\omega^2 LC$.

Row 18: This row computes the Gain using Rows 15 and 16.

Row 19: This row computes Gain in dB.

Row 20: This row computes the Phase in degrees.

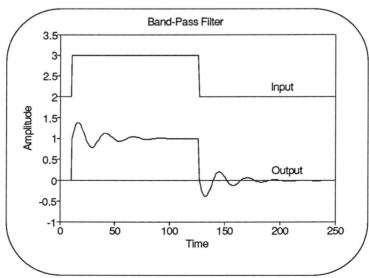

Fig. 7-2. Time domain response of a band-pass filter. Ringing depends on Q.

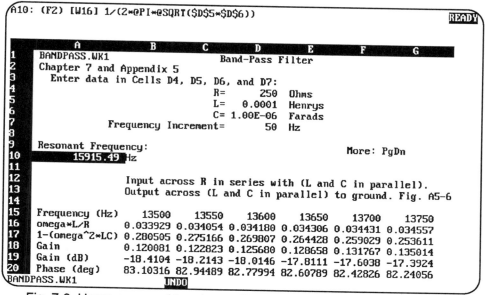

Fig. 7-3. Home screen of band-pass filter worksheet. Appendix 5 has analysis.

Press PgDn to see the second screen, which performs calculations that are normalized to the resonant frequency. The graphs connected to this worksheet are named *Gain*, *Gain [dB]*, *Gain [bar]*, and *Phase*; these are shown in Figures 7-4, 7-5, and 7-6, respectively.

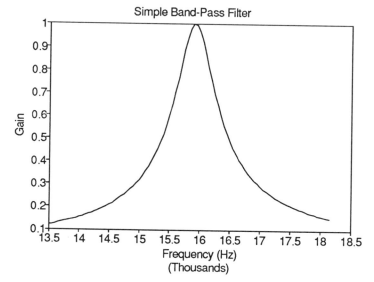

Fig. 7-4. Gain of a simple band-pass filter. Graph name: *Gain*.

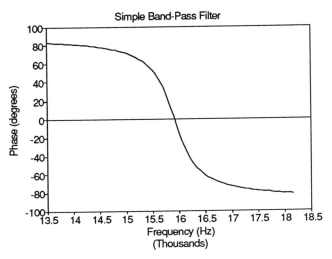

Fig. 7-5. Phase of a simple band-pass filter. The bandwidth is the frequency interval between the +45 and −45 degree points. Graph name: *Phase*.

Fig. 7-6 shows the gain of the band-pass filter on a bar graph, which emphasizes the fact that the calculations were done only at discrete frequencies. As a consequence of the discrete calculations the actual maximum gain of the filter may not be shown exactly. It will only be correct *if the gain was actually measured at the resonant frequency*.

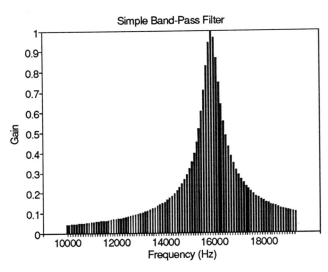

Fig. 7-6. Gain measured at discrete frequencies, linear on both axes.

Gain and Phase of Simple Band-Stop Filters

A band-stop, band-reject, or notch filter is used to minimize a narrow band of frequencies. The frequency response of a simple one is shown in Fig. 7-7.

BANDSTOP, Band-Stop Filter Worksheet

The Home screen of this worksheet is shown in Fig. 7-8. With this worksheet you can experiment with L, R, and C to produce single-pole band-stop filters. The circuit and its analysis is shown in Fig. A5-7 in Appendix 5.

Worksheet Organization

Cells D4, D5, D6, D7 and D8: Enter information on R, L, C, frequency increment, and internal resistance of L.

Cell A9: This cell computes and displays the resonant frequency.

Row 15: This row is the frequency axis. Enter the starting frequency in Cell B15.

Row 16: This row computes $2\pi fRC$ at each frequency.

Row 17: This row computes $1 - (2\pi f)^2 LC$.

Row 18: This row computes $2\pi f(R + r)C$.

Row 19: This row computes Gain using Rows 16, 17, and 18.

Row 20: This row computes Gain in dB.

Row 21: This row computes Phase in degrees. Press PgDn to see the second screen, which performs calculations normalized to the resonant (stop) frequency.

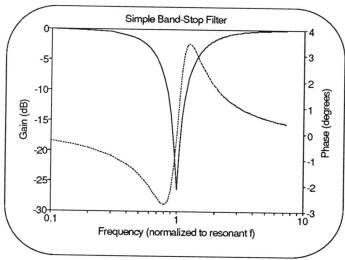

Fig. 7-7. Bode plot of a band-stop filter of the type shown in Fig. A5-7. The x-axis is normalized to the resonant (stop) frequency. Actual shapes depend on Q.

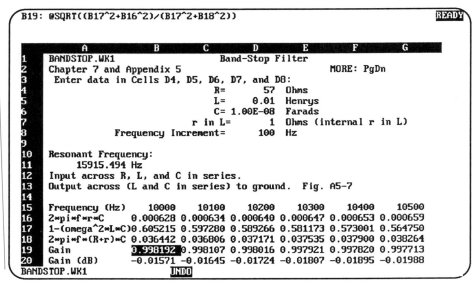

Fig. 7-8. Home screen of band-stop worksheet. Appendix 5 has analysis.

The Gain of the band-stop filter for the parameters in Fig. 7-8 is shown in Fig. 7-9. Figures 7-10, 7-11, and 7-12 show worksheet graphs *Gain [dB], Phase.,* and *Gain [dB bar].*

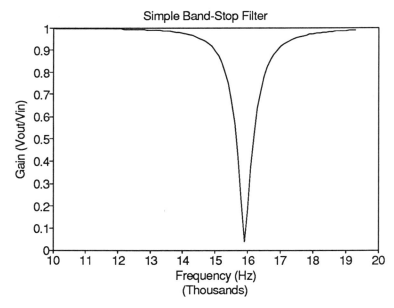

Fig. 7-9. Gain of band-stop filter on linear scales. Graph name: *Gain.*

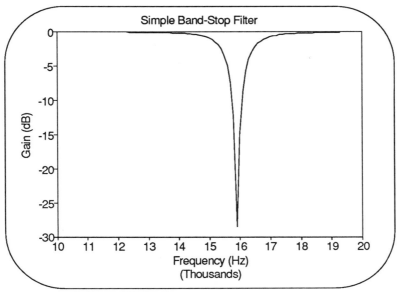

Fig. 7-10. Band-stop filter Gain in dB, with frequency on a linear scale. Graph name: *Gain [dB]*. These Figures are based on the parameters in Fig. 7-8.

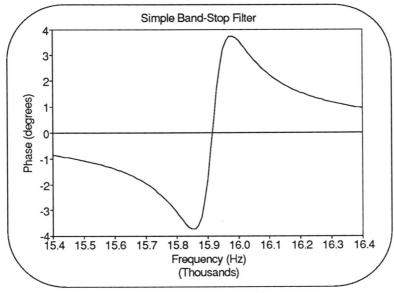

Fig. 7-11. Band-stop filter phase, on a linear frequency scale. Graph name: *Phase*.

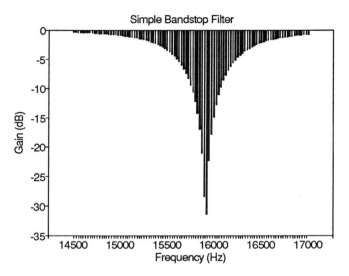

Fig. 7-12. Gain in dB measured at discrete frequencies, on a linear frequency scale. This is similar to the worksheet graph named *Gain [dB bar]*.

Standard Measurements of Hearing Aids

Hearing aids have more complicated Gain curves than the simple filters that we have discussed here. As with many amplifiers, hearing aids amplify a band of frequencies but they roll off the lows and the highs. Because hearing loss is mainly at the higher speech frequencies, hearing aids are designed to amplify the highs more than the lows. However, there is no need to amplify frequencies above those in the normal hearing range.

From a crude viewpoint a hearing aid can be regarded as an active band-pass filter. In addition, modern aids have features such as dynamic compression so that the Gain at a given frequency is not constant, but is a nonlinear function of the input (often a logarithmic function). Dynamic compression causes Gain to decrease as the input increases. The output transducer and its acoustic tube have a profound effect on the overall frequency response.

The accepted Standards for hearing aid measurements are updated as technology advances. A summary of typical definitions and measurements is in Appendix 3. In this section we give examples of applications of these specialized measurements.

Example of Calculation of High-Frequency Full-on Gain
Suppose you measure a hearing aid with full-on gain and input sound pressure level of 60 dB and obtain the results in Table 7-1.

Table 7-1. Data for High-Frequency Full-on Gain Example									
Freq. (Hz)	300	600	1000	1600	2000	2500	3000	3500	4000
Gain (dB)	22	53	56	77	83	86	77	53	35

What is the High-Frequency Full-on Gain?

This definition of gain is the average of the gains at 1000, 1600, and 2500 Hz, with volume control full-on and input sound pressure level of 60 dB. Therefore, for these data the High-Frequency Full-on Gain is

$$\frac{56 + 77 + 86}{3} = 73 \text{ dB}$$

Example of Frequency Response Measurement:

The frequency response is obtained by subtracting 20 dB from the High-Frequency Full-on Gain, then determining where this line intercepts the measured gain curve. For the data above, $73 - 20 = 53$ dB.

Plotting the data with the 53 dB line shows that the intercepts are at 600 Hz and 3500 Hz. The bandwidth can therefore be expressed as being from 600 Hz to 3500 Hz, or a bandwidth of 2900 Hz. This is shown in Fig. 7-13.

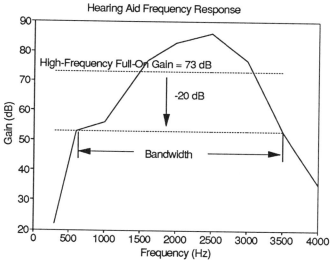

Fig. 7-13. Hearing aid frequency response measurement for data in Table 7-1. The High-Frequency Full-On Gain in this example is 73 dB. Drop down 20 dB, which gives the 53 dB horizontal line. The bandwidth is the frequency interval between the intercepts of the 53 dB line with the frequency response curve.

Shape Factor

The *shape factor* is a measure of the steepness of the slopes on the bandpass or band-stop frequency response curves. It is the ratio of the bandwidth at −60 dB to the bandwidth at −3 dB,

$$SF = \frac{-60 \text{ dB bandwidth}}{-3 \text{ dB bandwidth}} .$$

The smaller the shape factor, the steeper the slope and the better the skirt selectivity.

Insertion Loss

When a filter is inserted between a source and a load there is, inevitably, some power loss due to absorption and reflection. The sum of the absorption and reflection losses is called the insertion loss. Careful attention to impedance matching can minimize reflection loss and use of low-loss inductors and capacitors minimizes absorption. Any resistive component will contribute to absorption.

Ripple

The simple single-pole filters used in this Chapter have a single peak at the resonant frequency, but multipole filters designed for a broader response at the −3 dB points and more narrow response at the −60 dB points sacrifice some smoothness of the response curve in the pass or stop band. This lack of smoothness appears as multiple small peaks, like a ripple. The maximum peak-to-peak ripple, measured in dB, is commonly called "the ripple." Special filter designs are available that are maximally flat.

Ultimate Attenuation

For a passive bandpass filter the ultimate attenuation is the *minimum* attenuation outside of a specified passband. An ideal filter would provide infinite attenuation, but real filters always have some leakage.

Digging Deeper

DaDISP has a filter design module available as an add-in. DSP Development Corp., One Kendall Square, Cambridge, MA 02139.

Mathcad has a signal processing add-in that includes filter design. MathSoft, Inc., 101 Main Street, Cambridge, MA 02142.

A selection of filter design add-ins for spreadsheets is available from Engineering Solutions, P. O. Box 570159, Tarzana, CA 91356. Phone (818) 772-7231.

Eagleware has a full range of professional filter design and circuit simulation programs. These include *LC* filter, active *RC* filter, and microwave filter synthesis. Eagleware Corp., 1750 Mountain Glen, Stone Mountain, GA 30087. Phone (404) 939-0156; FAX (404) 939-0157.

Chapter 8

Frequency Domain Operations

What This Chapter is About

Viewing a signal in its spectral representation provides a different and often valuable new perspective. In this chapter we will explore some basic elements of the broad area of spectrum analysis. Several sophisticated operations can be performed very simply in the frequency domain using nothing more than multiplication and division. To enter the frequency domain some new versions of spreadsheets provide a Fast Fourier Transform (FFT) operation in their bag of tricks. For spreadsheets without this feature add-ins are available which perform the FFT, and practically every math program has FFT and/or Discrete Fourier Transform (DFT) capabilities.

A worksheet is provided here for Fourier series composition of periodic signals with and without non-periodic additive noise; your spreadsheet does not need the FFT operation for this. Other worksheets are provided for use with the FFT if this is available in your spreadsheet. If you do not have the FFT you can still see how the frequency domain operations perform in these worksheets, but you will not be able to make changes and see the results.

We will discuss a few of the subtleties of using the FFT and then apply it to computing the complex spectrum, its magnitude and phase representation, and the power spectrum. In Chapter 9 you'll learn how easy it is to perform filtering in the frequency domain. In Chapter 10 you will also see how to compute some other useful signal representations including the Cross-Spectrum, the Auto-Correlation and Cross-Correlation Functions, the Coherence Function, and the Transfer or System Function. We finish this brief tour of digital signal processing with an introduction to System Identification, which helps you to find out what's inside a system by putting a signal in and seeing what comes out.

132

Periodic Signals

Signals can be classified as periodic or non-periodic. This is like saying all living creatures can be divided into two classes, elephants and non-elephants. A periodic signal repeats itself after a finite interval. A periodic signal $s(t)$ turns into itself after its period T, so $s(t) = s(t + T)$. A non-periodic signal may occur only once, or at random intervals.

It is convenient for analysis to separate non-periodic signals further into transient signals (like lightning) and random signals (like noise). Many signals, such as speech, are mixtures of periodic and non-periodic components. Such a mixture is represented in Fig. 8-1, where a periodic signal seems to emerge as the signal/noise ratio improves.

Even the most complicated periodic signal can be represented by a Fourier series, under certain restrictions called the Dirichlet conditions. The Fourier series is simply a sum of sines and cosines of various amplitudes and frequencies. The Dirichlet conditions are not very restrictive:

- The signal must have a finite number of discontinuities in one period.

- The signal must have a finite number of maxima and minima in one period.

- The signal must be absolutely integrable over a period, i.e., the integral must be finite.

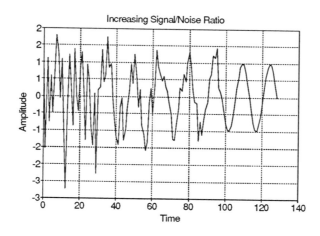

Fig. 8-1. A periodic signal appears to emerge as the signal/noise ratio increases. However, using only the information in these time-limited data it is possible that the entire signal is periodic, perhaps with a period of 128 time units. (Also see Fig. A2-1, Appendix 2.)

A fundamental problem in cryptography, radar, and military electronics is the determination (or estimation) of whether a signal is periodic or non-periodic. Some types of periodic signals may appear to be non-periodic (like noise) if they are observed for less than a full period. Modern electronics and computer technology can produce extraordinarily long sequences that are actually periodic but appear to be random if the observer lacks patience and/or the capability of analyzing these signals.

The importance of signal classification in terms of spectral analysis is:

- Periodic signals have *discrete* or *line* spectra.

- Non-periodic signals have *continuous* spectra.

Fourier Series

Fourier discovered an amazing and beautiful representation of periodic signals (subject to the restrictions of the Dirichlet conditions). He showed that the sine wave is the fundamental building block of Nature, insofar as any periodic signal can be constructed from them. The Fourier series can be expressed in several equivalent forms. One easily understood form represents a periodic function $s(t)$ in terms of an infinite series of sines (or cosines) of different amplitudes (A_n), frequencies (n), and phases (φ_v),

$$s(t) = \frac{A_o}{2} + \sum_{n=1}^{\infty} A_n \cos\left(\frac{2\pi nt}{T} + \varphi_n\right) \quad n = 1, 2, 3, \ldots, \infty$$

The term $A_o/2$ is the "dc" or average value of the signal. If the signal has an average value of zero then A_o is zero. The period is T.

For analytical computation an equivalent form involving both sines and cosines is more useful. This form explicitly decomposes the series into even (cosine) and odd (sine) parts, and may be written as,

$$s(t) = \frac{A_o}{2} + \sum_{n=1}^{\infty} A_n \cos\left(\frac{2\pi nt}{T}\right) + B_n \sin\left(\frac{2\pi nt}{T}\right), \quad n = 1, 2, 3, \ldots, \infty$$

The first few terms of the even part of the Fourier series look like,

$$\frac{A_o}{2} + A_1 \cos\left(\frac{2\pi \cdot 1 \cdot t}{T}\right) + A_2 \cos\left(\frac{2\pi \cdot 2 \cdot t}{T}\right) + \ldots$$

and the first two terms of the odd part looks like,

$$B_1 \sin\left(\frac{2\pi \cdot 1 \cdot t}{T}\right) + B_2 \sin\left(\frac{2\pi \cdot 2 \cdot t}{T}\right) + \ldots$$

A few remarks concerning the series may be helpful.

- A signal that is an even function will be represented entirely by the cosine series. An even function is one that turns into itself if it is folded about the time axis at t = 0. In other words, $s(t) = s(-t)$. An odd function, characterized by $s(t) = -s(-t)$, will be represented entirely by a sine series.

- The Fourier series can represent continuous *and* discontinuous functions. At points of discontinuity the series converges to the average value of the function at the discontinuity.

- When a function has a discontinuous slope the series representation has an overshoot called the Gibbs' effect. This overshoot decreases in importance as the number of terms in the series increases but it never disappears entirely. You can see this effect in the SYNSPECT worksheet described below.

- For a given number of terms, the Fourier series provides the optimum least-mean-squares (LMS) approximation of a periodic function.

For a pulse signal, the bandwidth is inversely proportional to the pulse width; the density of the line spectrum is inversely proportional to the pulse repetition frequency.

Fourier Analysis and Synthesis

Analysis is the process of decomposing a signal into its sine and cosine components, with the proper amplitudes and frequencies, that is, determining the values of the n's, A_n's, and B_n's. Synthesis is the process of combining various sines and cosines with the proper amplitudes and frequencies to produce a desired waveform.

Analysis usually requires methods of calculus or numerical integration which are complicated and tedious, but soon we will perform numerical Fourier analysis the easy way, using the FFT contained in spreadsheets. Synthesis is simple to do and is instructive because it shows how an infinite number of arbitrary periodic signals can be constructed using only sines and cosines.

SYNSPECT, Fourier Series Worksheet

This worksheet lets you synthesize a spectrum and its associated periodic waveform by choosing various harmonics and amplitudes. Figure 8-2 shows the Home screen. Table 8-1 has suggestions for the Fourier components of a square wave, a sawtooth wave, and a triangular wave. A square wave is a standard test signal because it is rich in harmonics. A sawtooth wave is typical of the horizontal sweep voltage in an oscilloscope or TV. Of these three, the triangular wave is closest to a sine wave. It is like a child might draw a sine wave with a ruler. In a triangular wave the harmonic amplitudes decrease as the inverse square of the harmonic number.

Table 8-1. Common Periodic Waveforms			
Signal Type	*Harmonic Number*	*A*	*B*
Square Wave	1	0	1/1
	3	0	1/3
	5	0	1/5
	(all odd harmonics)	(experiment with other values)	(reciprocal of odd harmonic number)
Sawtooth Wave	1	0	1/1
	2	0	−1/2
	3	0	1/3
	(all harmonics)	(experiment with other values)	(reciprocal of harmonic number; alternating sign)
Triangular Wave	1	0	$1/1^2$
	3	0	$−1/3^2$
	5	0	$1/5^2$
	(all odd harmonics)	(experiment with other values)	(reciprocal of square of odd harmonic number, alternating sign)

You can easily increase the number of Fourier components in the worksheet to experiment with reducing the ripple and ringing. You can also experiment with adding random noise to individual Fourier components. The worksheet displays the signal waveform in the graph named *Synthesizer* and the spectrum in the graphs named *Spectrum dB* and *Waveform*.

Worksheet Organization

Enter the A_n amplitudes in Cells C6...C15, the B_n amplitudes in Cells E6..E15, and the period in Cell G6. In setting the period T in Cell G6 be careful about undersampling. If the period is too short the higher-frequency Fourier components will have few data points per cycle and the results will be ragged.

Row 17 is the Time Axis. Use Fill operation to adjust the range and increment.

Rows 18..27 calculate the Fourier components at each time increment.

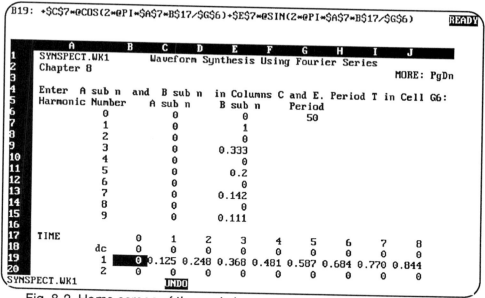

Fig. 8-2. Home screen of the worksheet SYNSPECT. The formula in Cell B19 is shown at the upper left of the screen.

The worksheet on the diskette has a time axis that goes from zero to 100, and the chosen period is 50. Experiment with periods shorter than 50 to see what happens with lower sampling. View the graph *Waveform* to check the results.

The results produced by the Fourier series are very sensitive to amplitudes and frequencies. You can easily experiment with the effects of errors in the synthesis by modifying the amplitudes in Cells C6..C15 and E6..E15. Make small changes in harmonic number in Cells A6..A15 and watch the results deteriorate as time goes by.

A square wave is a useful signal for testing systems because it is simple, easily produced and analyzed, and has a line spectrum with significant amplitudes at high harmonics because the amplitudes decrease inversely proportional to the first power of the harmonic number. When testing a system with a square wave you are testing simultaneously at many frequencies.

Viewing a square wave in the time domain (with an oscilloscope) gives you a quick look at the characteristics of a system. If the corners are rounded this indicates a high-frequency loss. If the top of the square wave sags this indicates a low-frequency loss. If there is ringing, like an exponentially damped oscillation, this indicates that the system has a relatively narrow bandwidth and a significant Q.

Figures 8-4 and 8-5 show the spectrum and waveform of a synthesized square wave. The spectrum and waveform of a synthesized sawtooth signal are shown in Figs. 8-5 and 8-6.

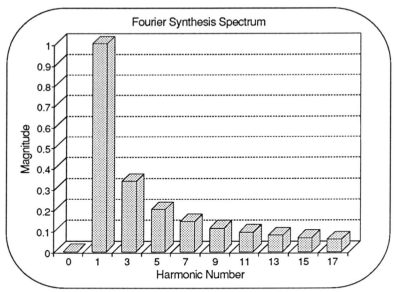

Fig. 8-3. Spectrum of the square wave synthesized in the worksheet shown in Fig. 8-2, with 9 odd harmonics. A perfect square wave has no even harmonics.

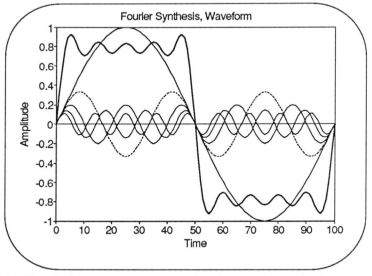

Fig. 8-4. This is the square wave synthesized in the worksheet shown in Fig. 8-2. When viewed in color on your monitor the role of the harmonics will be much clearer. Observe that there are 5 bumps on the square wave. This indicates the presence of 5 non-zero harmonics. This waveform has no dc component.

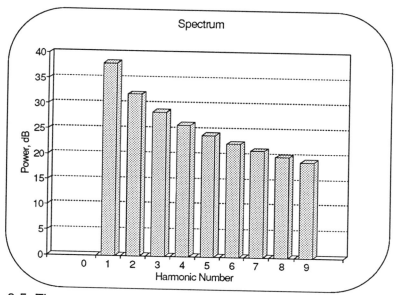

Fig. 8-5. The spectrum of a sawtooth signal (like the horizontal sweep of an oscilloscope) contains both even and odd harmonics. Nine harmonics are shown.

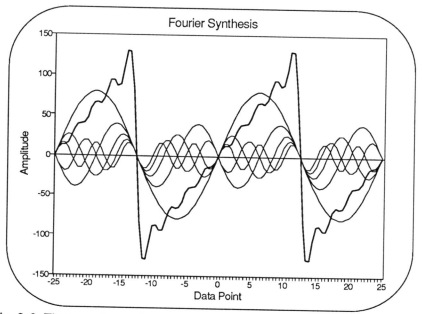

Fig. 8-6. The waveform of a sawtooth signal synthesized with 9 harmonics.

Viewing the square wave in the frequency domain can give very accurate quantitative results but *only at the frequencies of the harmonics*. For example, you can measure the amplitude of each harmonic at the input and at the output, and calculate the difference. This will give you a measure of the gain at each harmonic frequency. In order to measure the system characteristics at other frequencies you can use a sinewave input, but this is slower because it only measures at one frequency.

Non-Periodic Signals

Non-periodic signals have continuous spectra, which means that there is usually some energy at all frequencies. A non-periodic signal can often be represented in the frequency domain by a Fourier integral or by the Fast Fourier Tranform (FFT). I promised you in the Preface that you would not have to use calculus in this book, so we will enter the frequency domain with the FFT found in some of the newer versions of spreadsheets and most math programs.

Before describing how to operate with the FFT, let's briefly review the "@-functions" that will be useful. Because a real signal almost always results in a complex spectrum, we need to use some complex "@-functions" in our computations. Also, you need to learn a trick with the FFT that will produce results that look "normal" in the sense that they look like those that you see for spectra that can be calculated by hand.

Tips and Tricks for the FFT in Spreadsheets

Useful Operations and @-Functions

FFT. This operation (accessed with a mouse click) transforms a block of data from the time domain to the frequency domain using the Fast Fourier Transform. The number of cells must be an integer power of 2, like 32, 128, 1024, and so forth. Consult your spreadsheet manual to learn how to access this function. You must specify an input block and a destination or output block. Usually the output data will be complex. *Caution:* This function does not recalculate automatically. You must manually recalculate the FFT by accessing it or writing a macro to handle this task.

IFFT. This is the inverse Fast Fourier Transform, which converts from the frequency domain into the time domain. This function is accessed like the FFT, and requires specification of an input block and an output block. Even if you expect the output to be completely real, the output block (time domain data) will often have a small imaginary component due to round-off errors. You can recover the real part using the @IMREAL function. *Caution:* Like FFT, the IFFT does not recalculate automatically. *Tip:* Install an icon on the tool bar to activate Fourier Analysis.

@COMPLEX(,). This converts real and imaginary data into a complex number format. It will accept two numbers or two cell references.

@IMABS( ). This function computes the magnitude of a complex number, and it recalculates automatically when the complex number changes. This is useful for conversion of the rectangular form to polar form, together with the @IMARGU-MENT function.

@IMAGINARY( ). This function recovers the imaginary part of a complex number. You can copy it to a block of cells and it recalculates automatically when the complex number changes.

@IMARGUMENT( ). This function computes the phase angle *in radians* of a complex number, so it is useful for conversion from rectangular to polar form. To get the result in degrees use @DEGREES(@IMARGUMENT( )). This function recalculates when the complex number changes.

@IMCONJUGATE( ). This function computes the complex conjugate of a complex number, which is useful in several calculations discussed below. It recalculates when the complex number changes.

@IMCOS( ). This gives the cosine of a complex number.

@IMDIV(,). This returns the quotient of two complex numbers. This is useful in system identification and deconvolution. It accepts two complex numbers or two cell references.

@IMEXP( ). This gives the exponential of a complex number.

@IMLN( ). This gives the natural or base-e logarithm of a complex number.

@IMLOG10( ). This gives the common or base-10 logarithm of a complex number.

@IMLOG2( ). This gives the base-2 log of a complex number.

@IMPOWER(,). This raises a complex number to a complex power.

@IMPRODUCT(,). This multiplies two complex numbers; it is useful in filtering, convolution, and correlation.

@IMREAL( ). This function recovers the real part of a complex number. You can copy it to a block of cells and it recalculates automatically when the complex number changes.

@IMSIN( ). This gives the sine of a complex number.

@IMSQRT( ). This takes the square root of a complex number.

@IMSUB(,). This subtracts one complex number from another. It accepts complex numbers or cell references, and is useful in finding the difference between two spectra.

@IMSUM(,). This adds two complex numbers. It accepts complex numbers or cell references, and is useful in summing two spectra.

Centering Spectra: Nyquist Shift

The direct result of using the FFT gives spectra which may not look familiar because they are not centered. To see this effect we will compare an FFT spectrum with a spectrum for a simple case that can be calculated exactly. One effective method of centering an FFT spectrum is to use a technique called "checkerboarding". This consists of multiplying every odd data point in the original data domain by -1 before taking the FFT. This method, sometimes called a Nyquist shift, is the one that we will use in worksheets with the FFT.

Sampling and Aliasing

The FFT operates with digitized or sampled data and the amount of sampling has profound effect on the results. Nyquist showed that it is possible to reconstruct a continuous-time signal completely from discrete equally spaced samples, providing the highest frequency in the time signal is less than half the sampling frequency. Sampling at a lower rate results in "aliasing," which involves high frequencies masquerading as low frequencies. You can see this strobe effect in a movie (a two-dimensional sampled data system) when the wheels of vehicles appear to rotate backward or forward slowly even though the wheels are rapidly spinning.

The sampling frequency is often called the Nyquist frequency, and it must be at least twice the highest frequency that you can know with confidence. For example, if you sample at 100 kHz, the highest frequency that you can know with confidence is 50 kHz. However, it is much better to sample at several times the minimum rate, perhaps 8 or 16 times the minimum. Sometimes on compact disc players you see a label that says "8 times oversampling" which means that it samples eight times faster than the absolute minimum. This is much easier on filters. Be careful about undersampling; results deteriorate as you approach the Nyquist frequency. (See Fig. 8-29.)

To minimize effects of aliasing it is common practice to insert an "anti-aliasing filter" before a digital/analog converter. This is just a low-pass filter, used to reduce the high frequencies at the input. This filter must be designed to have a *linear phase* over the low frequencies of interest in order to minimize waveform distortion of the signal you want to digitize.

The Nyquist theorem provides the connection between continuous time signals and digitized signals, but it does not tell the complete story because the width of the sampling pulse and the start and stop of the sampling operation are also important. An abrupt start and stop of the measurement causes spectral leakage which smears frequency resolution (as we shall see shortly). Leakage involves the spreading of energy into adjacent frequencies; it can be reduced by proper choice of data windows.

Data Windowing

Subjecting a block of data to the FFT operation implicitly imposes a rectangular window on the data in the form of an abrupt start and stop. As we shall see, this results in loss of frequency resolution because of spectral leakage. By imposing another type of window to soften the start and stop of the data block you can make a trade-off in enhancing frequency resolution at the expense of amplitude resolution. This is a big subject and there are subtle issues that must be considered. Fortunately, with a spreadsheet it is easy to impose a variety of windows and exploit the trade-offs or empirically select one that gives acceptable results. The article by Harris and the books by Brigham and Marple listed at the end of this chapter discuss windowing in great detail, and we will pursue this further in Chapter 9.

Imposed Periodicity

The FFT assumes that the input data are periodic even if they are non-periodic. As a result, the FFT produces a discrete or line spectrum, not a continuous spectrum.

Zero Padding

Most FFT operations, like the ones found in spreadsheets, require a data set that consists of a number of points that is an integer power of 2. If your data do not fit this requirement you can satisfy it by padding the data set with zeros. Zero-padding has its good side and its bad side. The good side is that you can always get the proper number of points. The downside is that zero-padding decreases the variance and increases the bias. You should experiment on some signals with different amounts of zero-padding to get a feeling for its effects. We will see some examples later in this chapter.

Frequency Scaling

If Δt is the sampling interval in seconds then the sampling frequency is,

$$f_{sample} = \frac{1}{\Delta t} \ .$$

For example, if the interval between samples is 10^{-5} s, the sampling frequency is 100 kHz. Then, if the total number of samples in the data set is N, the frequency scale is,

$$f = \frac{1}{N\Delta t} \ .$$

To see how this works, suppose N is 512 and Δt is 10 μs (corresponding to a sampling rate of 100 kHz, commonly available in data acquisition boards). At 100 kHz sampling rate, according to the Nyquist sampling theorem, the highest frequency that you can know with confidence is 50 kHz. With these parameters the frequency interval between points is,

$$\frac{10^5}{512} = 195.3125 \text{ Hz}.$$

This is the smallest frequency interval that you can measure under these conditions. The only frequencies that you can measure with this are integer multiples of this basic increment. Now it is easy to see the importance of acquiring many samples (or, less desirable, increasing the sampling interval) to achieve better frequency resolution.

Complex Spectrum

As a first example of the FFT operation, let's use an exponential decay for the input because it is simple and its Fourier transform can be calculated exactly for comparison purposes. The complex spectrum can be expressed in Cartesian or rectangular form as,

$$S(f) = \text{Re}(f) + j\text{Im}(f)$$

where $\text{Re}(f)$ is the real part, j is the symbol that marks the imaginary part, and $\text{Im}(f)$ is the imaginary part (see Appendix 4). The equivalent expression in polar form is,

$$S(f) = M(f) \, exp[j\theta(f)]$$

where $M(f)$ is the magnitude and $\theta(f)$ is the phase.

FFT-XMPL, Fast Fourier Transform Example Worksheet

This worksheet provides an introduction to the use of the FFT operation in a spreadsheet. Figure 8-7 shows the Home screen. These are the operations that we will perform:

- First we will use the FFT to compute the complex spectrum. Direct use of the FFT produces a funny-looking result because the spectrum is not centered.
- Second, we will use a Nyquist shift (checkerboard) to center the spectrum so it will look familiar, or at least normal.
- Third, we will use @IMREAL and @IMAGINARY to select the real and imaginary parts of the spectrum.
- Fourth, we will use @IMABS and @IMARGUMENT to recover the magnitude and phase.
- Fifth, we will square the magnitude to get the power spectrum.
- Finally, we will compare the FFT results with an exact calculation.

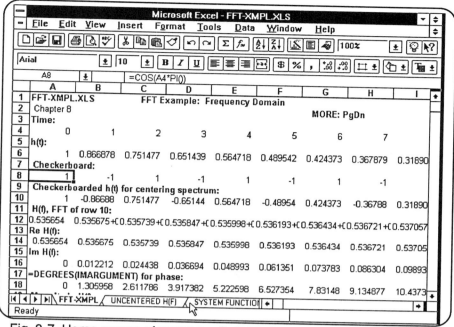

Fig. 8-7. Home screen of FFT-XMPL. Cell A8 shows the checkerboard formula.

Worksheet Organization

Row 4 is the Time axis. Use the Fill command to set the range and increment. The number of cells in the range must be an integer power of 2, such as 64, 128, 256, and so forth. For a large data set use a column format instead of a row format to obtain a maximum of 8192 data points.

Row 6 is the input signal. If the signal does not occupy a number of data points equal to an integer power of 2 then pad the signal end with zeros to reach the next power of 2. for example, if the signal has 58 points then add 6 zeros to obtain 64 points. Figure 8-8 shows the currently installed input signal.

Row 8 is the checkerboard (alternating +1 and –1) used to center the spectrum.

Row 10 performs the checkerboarding by multiplying each Cell in the signal by the checkerboard value the Cell column. For example, the formula in Cell A10 is +A6*A8.

Row 12 is the complex spectrum of Row 10. (To obtain an uncentered spectrum use Row 6 as the input.) To obtain Row 12 click on Fourier Analysis and a dialog box will appear on your screen. Type in the Input Row and the Output Row where requested. Both rows must have a number of cells that are an integer power of 2. Notice that there is a small square that you can click on to obtain the inverse FFT. Do not click on the inverse FFT. *Caution:* Each time the input signal changes you must activate Fourier Analysis and perform the FFT; it does not recalculate automatically. (You could write a macro specific to your spreadsheet to do this.) Observe that the values in Row 12 are generally complex.

Row 14 retrieves the real part of the complex spectrum of Row 12. The formula in Cell A14 is @IMREAL(A12).

Row 16 retrieves the imaginary part of the complex spectrum of Row 12. The formula in Cell A16 is @IMAGINARY(A12).

Row 18 retrieves the phase (in degrees) of the complex spectrum of Row 12. The formula in Cell A18 is @DEGREES(@IMARGUMENT(A12)).

Row 20 retrieves the magnitude of the complex spectrum of Row 12. The formula in Cell A20 is @IMABS(A12).

Row 24 scales the magnitude of Row 20 in dB. The formula in Cell A24 is 20*@LOG(A20).

Rows 26-34 are for tutorial purposes. They compute the spectrum of an exponential decay exactly for comparison with the FFT results.

Row 26 is the Frequency axis for the exact complex spectrum.

Row 28 is the real part of the exact spectrum. Place the cursor on Cell A28 to see its formula.

Row 30 is the imaginary part of the exact spectrum. Place the cursor on Cell A30 to see its formula.

Row 32 is the magnitude of the exact spectrum. The formula in Cell A32 is @SQRT(A28^2+A30^2).

Row 34 is the phase of the exact spectrum. The formula in Cell A34 is @DEGREES(@ATAN2(A28,A30)).

Rows 37- 41 compute the uncentered complex spectrum of Row 6, and the real and imaginary parts. A typical input signal is shown in Fig. 8-8 and its spectrum is shown in Figs. 8-9 - 8-11. Input any signal and FFT-XMPL will process its spectrum.

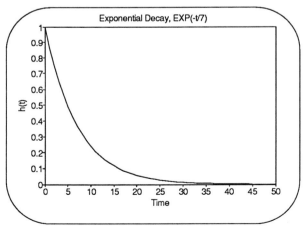

Fig. 8-8. This exponential decay is a transient event like the ones we encountered in Chapter 3. This is the input for FFT-XMPL. Graph name: *Input.*

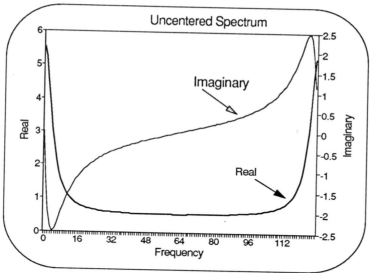

Fig. 8-9. This is an equivalent frequency-domain representation of the exponential decay shown in the time domain in Fig. 8-8. This was obtained by applying the FFT directly to the data in Fig. 8-8, which produces an uncentered spectrum. Graph name: *Uncentered FFT*.

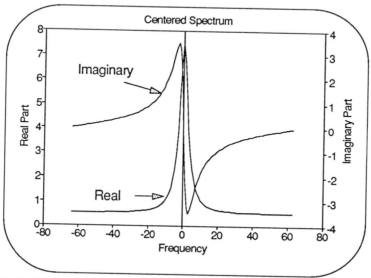

Fig. 8-10. This is another equivalent frequency-domain representation of the exponential decay in Fig. 8-8. Here the spectrum is centered by applying a Nyquist shift (checkerboard) before the FFT. Compare with Fig. 8-9. Graph name: *FFT Re&Im*.

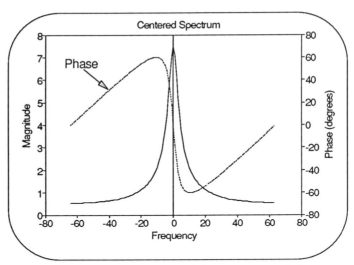

Fig. 8-11. The magnitude and phase of the spectrum of the exponential decay is shown in centered form. Graph name: *FFT M&P*.

The FFT gives good results close to the center frequency but the circular property produces errors outside of the half-bandwidth. Compare the previous Figures with Figs. 8-12 and 8-13 which were computed analytically for an exponential decay.

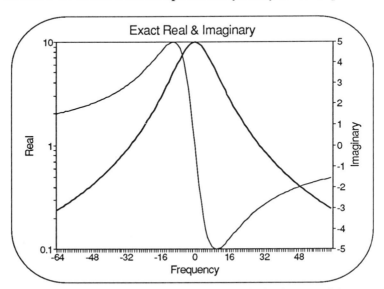

Fig. 8-12. The real (thick line) and imaginary (thin line) parts of the exact complex spectrum of a single exponential decay event. The results here are exact. Compare with Fig. 8-10 and note the error at the ends of the spectrum, due to the fact that the FFT imposes periodicity. Graph name: *Exact Re&Im*.

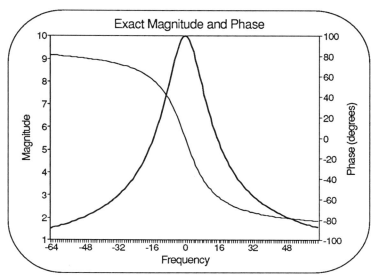

Fig. 8-13. The magnitude (thick line) and phase (thin line) of the exact complex spectrum of a single exponential decay event. Compare the phase error of the FFT at the ends of the spectrum in Fig. 8-11. Graph name: *Exact M&P*.

Polar Plots

A different insight into the nature of a spectrum can be obtained by a polar plot of the imaginary part of the spectrum vs. the real part, with frequency as a parameter. For the exponential decay spectrum the polar plot shows that the phase continuously decreases as the magnitude shrinks. The polar plot also provides a dramatic comparison between the exact result of a single decay and the FFT result, showing how the FFT produces closure or circularity because of its assumption of periodicity.

Figures 8-14 shows the polar plot of the exact complex spectrum. Compare this with Fig. 8-15 for the FFT. Closure is a consequence of circularity in the FFT, due to imposed periodicity where none may exist. The FFT is not reliable near the end points.

Power Spectrum

The power spectrum is an important signal description because it tells you *the power at each frequency*. Among other things, the power spectrum is handy in analyzing system nonlinearity in terms of harmonic generation and intermodulation distortion. The power spectrum is a completely real function and is defined as,

$$G_{1,1}(f) = S_1^*(f)\, S_1(f).$$

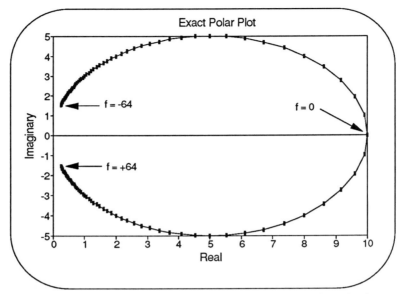

Fig. 8-14. Polar plot of the exact complex spectrum of a single exponential decay. The real part is the x-axis series and the imaginary part is the y-axis series. Frequency interval between markers is constant. Closure on the left would be at ±∞ frequency. Graph name: *Exact Polar*.

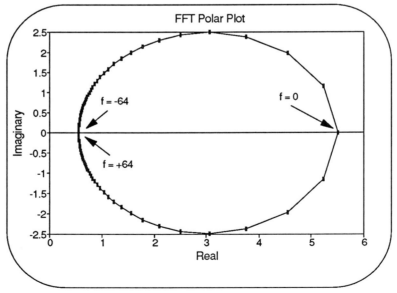

Fig. 8-15. Polar plot of the FFT of a single exponential decay. Observe closure (circularity) at the extremes of frequency, ±64. Graph name: *FFT Polar*.

In the definition of the the power spectrum $S_1^*(f)$ is the complex conjugate of the complex spectrum $S_1(f)$. The subscripts on the functions are not really necessary here, but they are used to emphasize the fact that everything belongs to one signal. In Chapter 9 we will use spectra from two different signals. The power spectrum for the exponential decay of Fig. 8-8 is shown in Fig. 8-16.

Pulse Spectrum

Finally, let's examine the spectrum of what is probably the most common signal today, a rectangular pulse. Both the exponential decay and the rectangular pulse are well-behaved signals that have Fourier transforms that can be calculated exactly by simple calculus, so they are useful for comparison with FFT results.

Several parameters are used to characterise a rectangular pulse. The rise and fall times of real pulses are not infinitely fast because all bandwidths are eventually limited. A deficit in low-frequency components may cause sag or droop of the top of the pulse.

The two parameters that we are most concerned with here are the pulse width and the pulse repetition frequency (PRF) because they have profoundly important effects on spectra. The PRF is the reciprocal of the time interval between the start of successive pulses in a pulse train. For example, if the time interval is one millisecond then the PRF is one kHz. As we shall see, the bandwidth of a pulse signal is the reciprocal of the pulse width. For example, a pulse width of one microsecond has a bandwidth of one MHz.

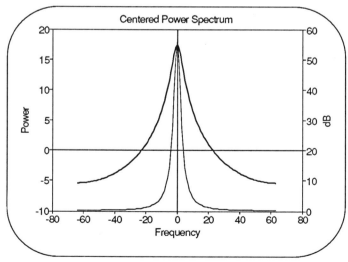

Fig. 8-16. Power spectrum of an exponential decay. There are no zeros in this spectrum, indicating that there is some power at every frequency. Thin line, left *y*-axis. Thick line, right *y*-axis. Graph Name: *FFT Power*.

FFTPULSE, Pulse Spectrum Worksheet

This worksheet lets you experiment with different pulses and pulse groups to investigate the effects on their spectra. It also includes an exact spectrum of a rectangular pulse.

Worksheet Organization

When you bring the Home screen of this worksheet up on your computer you will see the following rows:

Row 4: Time axis. Use the Fill command to set range and increment. The total number of data points should be an integer power of 2, like 128, 256, etc.

Row 6: Input pulse. As supplied on the disk, the input pulse starts at $t = 0$ and is 7 time units long. Figure 8-17 shows this pulse.

Row 8: Checkerboard (Nyquist shift) to center the spectrum. The formula in Cell A8 is @cos(A4*@PI), which is a convenient way of generating the checkerboard. This formula is copied to the other Cells in this row.

Row 10: Checkerboarded input. The formula in Cell A10 is +A6*A8. This formula is copied to the other Cells in this row.

Row 12: This is the complex spectrum returned by the FFT. When you execute the FFT a dialog box will appear. In the dialog box, set the input block (for 128 data points) to A10..DX10. The corresponding output block should be A12..DX12.

Row 14: The formula in this Cell is @IMREAL(A12), to obtain the real part of the spectrum. This formula is copied to the rest of this row. See Fig. 8-18.

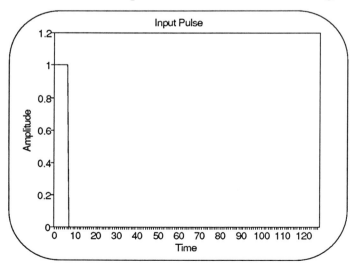

Fig. 8-17. This is the input pulse in the time domain. You can experiment with different pulse lengths, pulse shapes, and several pulses spaced at various intervals. This data set has 128 points.

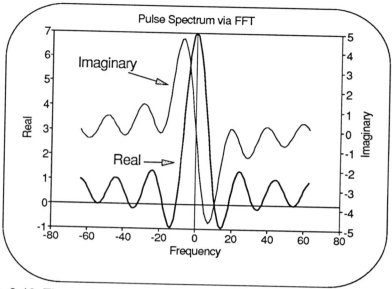

Fig. 8-18. The centered complex spectrum of the pulse shown in Fig. 8-17, as computed by the FFT. Observe that the imaginary part seems to be drifting up as frequency goes from left to right.

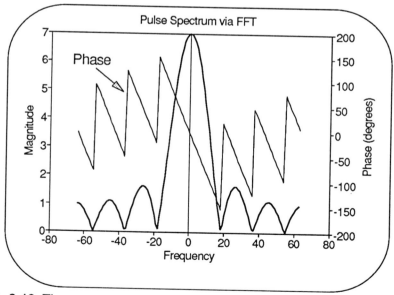

Fig. 8-19. The spectrum of the pulse shown in Fig. 8-17, in terms of magnitude and phase. Observe that the phase (thin line) appears to be drifting and the magnitude does not go all the way to zero on both sides of the central lobe.

Row 16: The formula in Cell A12 is @IMAGINARY(A12). This formula is copied to the rest of this row. The FFT polar plot is shown in Fig. 8-21.

Row 18: This row computes the phase of spectrum. The formula in Cell A18 is @DEGREES(@IMARGUMENT(A12)). This formula is copied to the rest of this row. See Fig. 8-19.

Row 20: Magnitude of spectrum. The formula in Cell A20 is @IMABS(A12)). This formula is copied to the rest of this row. See Fig. 8-19.

Row 22: Power spectral density. The formula in Cell A22 is +A20^2. This formula is copied to the rest of this row. See Fig. 8-20, left y-axis.

Row 24: Power spectral density in dB. The formula contained in Cell A24 is 10*@LOG(A22). This formula is copied to the rest of this row. See Fig. 8-20.

Row 26: Frequency axis for the exact spectrum. Use the Fill command to set the range and increment. As supplied on the disk, the range is –64 to +64 in unit steps.

Row 28: Real part of the exact spectrum. The formula contained in Cell A28 is @SIN(0.25*A26)/(0.25*A26). This formula is copied to the rest of this row. See Fig. 8-22.

Row 30: Imaginary part of the exact spectrum. The formula in Cell A30 is (@COS(0.25*A26)-1)/(0.25*A26). This formula is copied to the rest of this row.

Row 32: Magnitude of the exact spectrum. The formula contained in Cell A30 is @SQRT(A28^2 +A30^2). This formula is copied to the rest of this row.

Row 34: Phase of the exact spectrum. The formula contained in Cell A34 is @DEGREES(@ATAN2(A28,A30)). This formula is copied to the rest of this row.

Row 36: Power spectrum of the exact spectrum. The formula in Cell A36 is (A28^2 +A30^2). This formula is copied to the rest of this row. See Fig. 8-24.

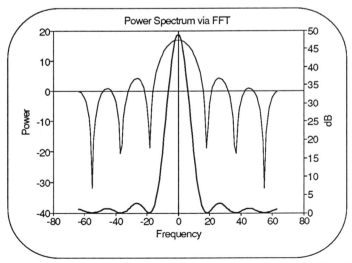

Fig. 8-20. Power spectrum of the pulse shown in Fig. 8-17. The thick curve is associated with the left axis; the thin curve is in dB.

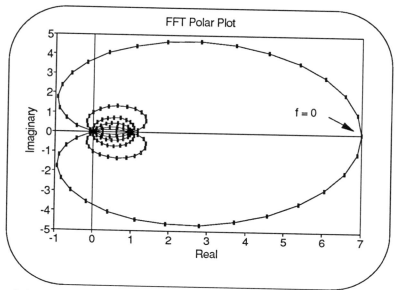

Fig. 8-21. Polar plot of the spectrum of the pulse shown in Fig. 8-17. The upper branch of the curve is the negative frequency domain; the lower branch belongs to the positive frequencies. Compare with Fig. 8-15 for the exponential decay.

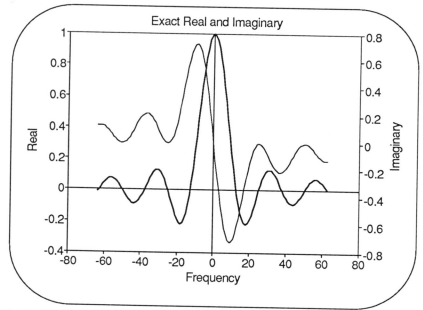

Fig. 8-22. The exact complex spectrum for a rectangular pulse similar to the one shown in Fig. 8-17. Thick line: Real. Compare with Fig. 8-18.

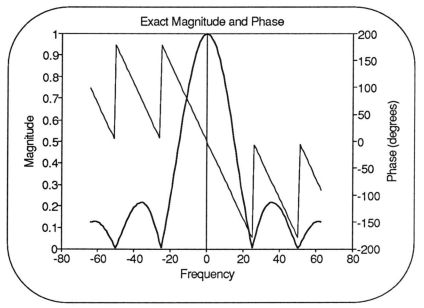

Fig. 8-23. Magnitude and phase of the exact spectrum for the rectangular pulse. Thick line: Magnitude. Compare with the FFT results shown in Fig. 8-19.

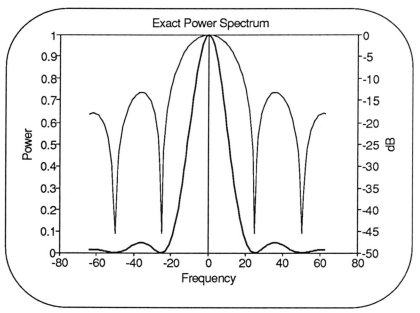

Fig. 8-24. Power spectrum of the exact complex spectrum for the rectangular pulse. Thin line: dB. Compare with the FFT results shown in Fig. 8-20.

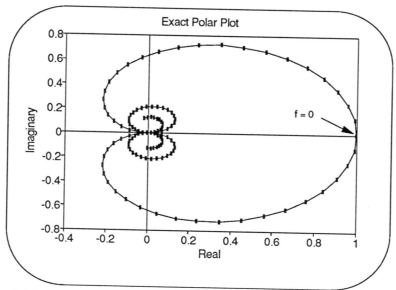

Fig. 8-25. Polar plot of the exact complex spectrum for the rectangular pulse. Frequency interval between markers is constant. Compare with the FFT results shown in Fig. 8-21. Graph name: *Exact Polar.*

Row 38: Power spectrum of the exact spectrum in dB. The formula in Cell A38 is 10*@LOG(A36). This formula is copied to the rest of this row.

Row 40: Uncentered input spectrum. This is obtained by choosing the input block for the FFT to be A6..DX6. This block is the original, un-checkerboarded input.

It is interesting to compare the FFT results for the rectangular pulse with the results of an exact calculation, as we did for the exponential decay. Again you will see the errors increasing as the frequency approaches the ends because of the imposed periodicity of the FFT. This particularly apparent in the phase. Note the apparent drift upwards in the phase in Fig. 8-19 co;mpared with Fig. 8-23.

Testing FFT and IFFT Operation

How good are the FFT and IFFT in your spreadsheet? One way to check them is to input any test signal as a waveform in the time domain and perform the FFT, obtaining the complex spectrum. Then perform the IFFT on this spectrum. If the operations were perfect you would expect to get out the original test signal in the time domain. A real input signal before the FFT should be *completely real* after the IFFT. However, usually due to round-off and other errors the output of the IFFT will be *complex* even when the original input is purely real. The errors are small but may be significant in some situations and you should be aware that they exist. For one thing, you will probably have to use @IMREAL to get the real part of the results of the IFFT. Test your spreadsheet!

Effects of Zero-Padding

We mentioned earlier that most FFT operations require the data set to have a number of points equal to an integer power of 2, like 32, 64, 128, 256, etc. Let's see the effects of padding a data set with zeros to achieve this requirement. For this test we'll use a rectangular pulse of unit amplitude and seven time units in length, starting at t = 0. We'll make three tests to bring the pulse sequence to lengths of 16, 32, and 64. The thick line is the magnitude and the thin line is the phase. The center is zero frequency. See Figs. 8-26, 8-27, and 8-28.

Unresolved Issues

Digitizing Errors

It is important to distinguish between *accuracy* and *precision*. Accuracy refers to the error of a measurement referred to some standard. For example, if something has a value 6.5 and you measure it as 6.45, then the accuracy is (6.5 – 6.45)/6.5 or approximately 0.77%. Usually a sequence of measurements of the same quantity will produce a *distribution* of results that will more or less peak around a central value. The spread of these measurements is the precision. A common measure of spread or precision is the standard deviation or its square, the variance.

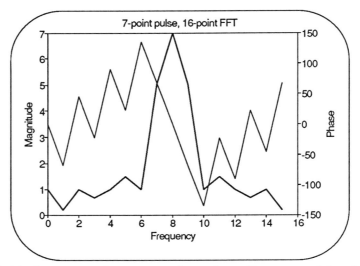

Fig. 8-26. Spectrum of a rectangular pulse with zero-padding to bring the sequence to a length of 16. The magnitude is ragged and the phase drifts up.

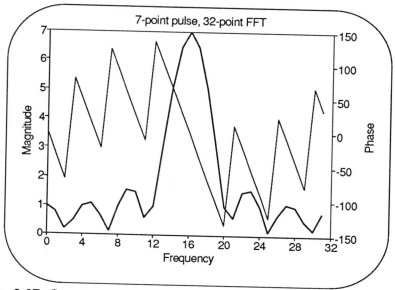

Fig. 8-27. Spectrum of a rectangular pulse with zero-padding to bring the sequence to a length of 32. The spectrum begins to look more normal.

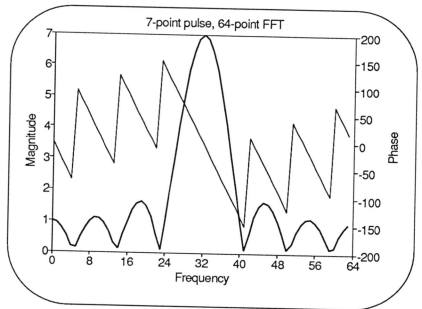

Fig. 8-28. Spectrum of a rectangular pulse with zero-padding to bring the sequence to a length of 64. Things are looking much better. Compare with Fig. 8-19 for a length of 128.

Sampling and digitizing entail additional considerations. For example, let's examine a measurement of a digitized sinewave. The result of the measurement will be a data set of discrete voltages at discrete time intervals. Here we have to distinguish between horizontal and vertical resolution. As we showed earlier, the time-sampling process produces a data set with a resolution that depends on the sampling rate $1/\Delta t$, where Δt is the sampling interval. For example, 1,000 samples per second gives a resolution of one millisecond, but 100 samples per second has a resolution of only 10 milliseconds. You really don't know what happens between samples. See Fig. 8-29.

The vertical resolution (of a noiseless signal) is limited by the number of bits in the A/D converter. A 12-bit digitization will divide the vertical axis into 2^{12} or 4,096 parts, so a sine wave with an amplitude of 10 Volts will be divided into steps of 2.441 mV, which corresponds to an error of 0.024%. If you measure the sinewave at a time when its amplitude is not exactly equal to one of these steps then a quantization error occurs. This has the appearance of noise added to the sine wave.

To further complicate matters the original signal may have noise. Let's face it, it *will* have noise. Electrical noise is an inevitable consequence of the motion of electrons (see Chapter 11). If the signal/noise ratio is low the precision will be low. There may also be non-random interfering signals that need to be removed or at least reduced.

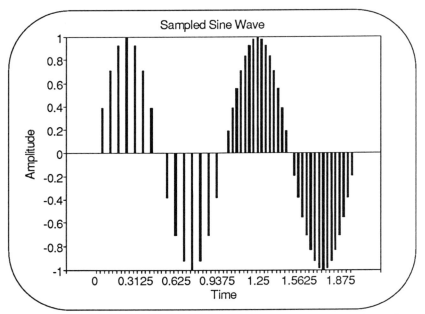

Fig. 8-29. Sampling rate and number of bits in A/D converter introduce errors in a digitized data set representing a sinewave.

The sampling rate and the digitization bits are certainly fundamental limitations but they can be minimized with high-quality hardware. The randomness (noise) of the desired signal can often be reduced by use of filtering, smoothing, and ensemble averaging techniques.

Spectrum Leakage

We have already mentioned the problem of aliasing, which makes high frequencies appear as low frequencies, and we briefly discussed data windowing to reduce leakage among adjacent spectral lines. Spectral resolution can be enhanced by reducing the acquisition of frequencies above the Nyquist frequency and by judicious selection of data windows. Examples and a worksheet for resolution enhancement by windows will be taken up in Chapter 9.

The width of the data acquisition window also affects spectrum leakage. You can see this by experimenting with the rectangular pulse and the FFT. The shorter the pulse, the wider the bandwidth. Spectrum leakage can be reduced by using a wider window but this in turn reduces time resolution because with a long window the signal may be changing while you are acquiring it. That's life, and you have to make trade-offs if you can. Joint time-frequency analysis (JTFA) may help. See Appendix 2. You have to do the best you can with what you have to work with, but you should be aware of errors and never treat experimental data as a sacred text.

Power Spectrum Normalization

There is no general agreement among analysts about what we have called the power spectrum. It is frequently called the power spectrum, the power density, the spectral density, the power spectral density (PSD), or the power density spectrum (PDS). There is also no agreement about normalization. One popular normalization is to divide the power spectrum by the total number of data points. (See Parseval's theorem, Appendix 2.) We are leaving the power spectrum in its "bare" form, and you can normalize it as you wish.

Brigham says, "The measurement of power spectra is a difficult and often misunderstood topic." Results using the FFT can be most successfully interpreted for signals that are periodic or deterministic. Random signals with the FFT is a subject of current research involving statistical estimation.

What's Next?

You now have the FFT, the key to the front door of the Taj Mahal, and you've seen a little of the beauty and mystery inside. Next we will explore some of the secret passages where the treasures are stored. In Chapter 9 we will look at some single-channel applications where only one input is present. Chapter 10 will guide our ascent into the labyrinth of multi-channel analysis.

Digging Deeper

The Fast Fourier Transform and Its Applications, E. O. Brigham (Prentice-Hall, Englewood Cliffs, New Jersey, 1988). ISBN 0-13-307505-2. Almost all you need to know.

Programs for Digital Signal Processing, Digital Signal Processing Committee, Ed. (IEEE Press and John Wiley & Sons, New York, 1979). ISBN 0-471-05962-5.

F. J. Harris, "On the use of windows for harmonic analysis with the discrete Fourier transform," *Proceedings of the IEEE*, **66**, pp. 51 - 83 (1978). This is a very valuable, very clear, often cited paper that tabulates and graphically compares 23 different windows. It also treats some common errors when windows are used with the FFT.

Modern Spectrum Analysis, II, S. B. Kesler, Ed. (IEEE Press, New York, 1986) ISBN 0-87942-203-3. A good companion to read with the books by Brigham and Marple.

Digital Spectral Analysis with Applications, S. L. Marple, Jr. (Prentice-Hall, Englewood Cliffs, New Jersey, 1987) .ISBN 0-13-214149-3. A diskette is included.

SSP, The Spreadsheet Signal Processor, S. C. Bloch (Prentice-Hall, Englewood Cliffs, NJ, 1992). ISBN 0-13-830506-4. Chapter 9 shows how to implement the Discrete Fourier Transform in a spreadsheet. Included are two diskettes with a variety of other signal processing worksheets.

Quattro Pro for Scientific and Engineering Spreadsheets, R. G. Parks (Springer-Verlag, New York, 1992). ISBN 0-387-97636-1. There is no signal processing but there are practical examples which specifically illustrate *Quattro Pro*'s capabilities. This book is aimed at Version 3.0 for DOS and is dated in its spreadsheet information but the applications are timeless.

1-2-3 for Scientists & Engineers, W. J. Orvis (SYBEX Inc., Alameda, California, 1987). ISBN 0-89588-407-0. This book is based on Lotus *1-2-3*, Version 2.01 for DOS. Spreadsheets have greatly improved since those days, but a wide variety of useful applications is described here.

Excel (Microsoft, Inc.) is a popular spreadsheet for Windows and the Macintosh. Starting with version 4.0 *Excel* has an FFT and IFFT in its Analysis Toolpak. Many data acquisition systems, like *LabView* and *LabWindows* (National Instruments) and *Labtech Notebook* (Labtech), can transfer data seamlessly into spreadsheets.

Quattro Pro (WordPerfect, Novell Applications Group) is another powerful spread-sheet with Windows and DOS versions. Starting with Version 5.0, *Quattro Pro* has an FFT and IFFT in its Analysis Tools. As in other spreadsheets, files can be opened and saved in a variety of *Quattro Pro*, *Excel*, and Lotus *1-2-3* formats.

An inexpensive FFT add-in for the *Excel* spreadsheet is available from Engineering Solutions, P. O. Box 570159, Tarzana, California 91356. This add-in will handle a single data set up to 1600 points or two data sets of up to 800 points each. It has several convenient and desirable features not found in *Excel's* standard FFT. As examples, it calculates the spectrum and cross-spectrum, and has provision for low-pass and high-pass filtering and von Hann windowing. It also automatically pads the input data with zeros to provide a number of points that is an integer power of 2.

FFTools is a useful FFT spreadsheet add-in. (DH Systems, Inc., 1940 Cotner Avenue, Los Angeles, California 90025.) Versions are available for Microsoft *Excel* and Lotus *1-2-3*. This add-in will handle data sets up to 8192 points (in integer powers of 2) and has 6 windowing functions (see Chapter 9).

Prime Factor FFT is a stand-alone program that operates with any Windows or DOS compiler. Under Windows it operates as a dynamic link library (or DLL). This program provides a choice of the standard Cooley-Tukey FFT, the *Prime Factor FFT*, and a DFT. *Prime Factor FFT* is not limited to powers of 2 so zero-padding is not needed. It operates on both one-dimensional and two-dimensional arrays (square or rectangular) which is useful for image processing. Alligator Technologies, 2900 Bristol Street, Suite E-101, Costa Mesa, CA, 92626-7906. Phone (714) 850-9984; FAX (714) 850-9987.

DADiSP, a "graphical spreadsheet," includes the FFT and DFT operations. Additional software is available for use with a variety of data acquisition boards so that real-world data can be loaded directly into *DADiSP*. DSP Development Corp., One Kendall Square, Cambridge, MA 02139.

Mathcad, a full-featured technical and scientific math program, has a built-in FFT operation. (MathSoft, Inc., 101 Main Street, Cambridge, Massachusetts 02142.)

LabVIEW Student Edition User's Guide, Lisa K. Wells, (Prentice-Hall, Englewood Cliffs, NJ, 1994). ISBN 0-13-210683-3. This book and its four diskettes provide an entry to laboratory and industrial-grade data acquisition and signal processing. Software includes the FFT and many signal processing and statistical operations. It will output a spreadsheet file with columns separated by tabs and rows separated by carriage returns. Available for the PC and the Mac; 6 MB required and

8 MB is recommended. Requires data acquisition hardware compatible with National Instruments boards. The Student Edition will make you want to upgrade to the Professional Version.

Chapter 9

Advanced FFT Operations, I

What This Chapter Is About

In Chapter 8 we explored a few basic operations involved in transforming into the frequency domain with the FFT. Using the IFFT we found that we could return to the original data domain. With that preparation we are ready to plunge more deeply into the frequency domain and learn some more sophisticated tricks.

In this chapter we will examine four single-channel operations. This means that the system has only one input and one output. You can think of the system's input and output in terms of the waveforms in the time domain or the spectra in the frequency domain shown in Fig. 9-1.

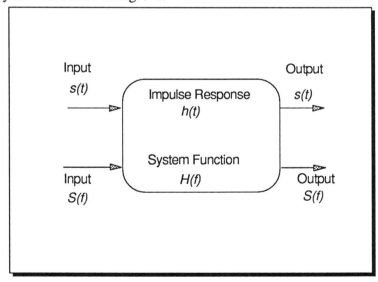

Fig. 9-1. A system can be described equivalently in terms of the time domain or the frequency domain. In the time domain *h(t)* determines the output waveform *s(t)*; in the frequency domain *H(f)* determines the output spectrum *S(f)*.

In this chapter worksheets are provided for:

- System function (Transfer Function) and Impulse Response
- Convolution
- Deconvolution
- Data windows
- Frequency resolution tests

System Function, Transfer Function

What is it?

The impulse response $h(t)$ and the system function $H(f)$, sometimes called transfer function, comprise a Fourier transform pair,

$$H(f) = \text{FFT } h(t)$$

$$h(t) = \text{IFFT } H(f)$$

FFT is the Fast Fourier Transform, and IFFT is the Inverse FFT. Each one of the pair contains the *complete* information about the system, because if you know one you know the other by means of the Fourier transform. For example, in a movie theater a swept-frequency or "chirp" signal can be used to measure the system function $H(f)$, and then the IFFT is used to obtain the impulse response $h(t)$.

This can measure the entire "B-chain" which consists of the room and loud-speaker equalization, power amplifiers, cross-overs and loudspeakers, screen loss, and auditorium acoustics (see Digging Deeper at the end of this chapter). At a simpler level, you could measure the characteristics of a filter by measuring $H(f)$ with a swept frequency, or a sine wave at discrete frequencies. Or, you could use a very brief pulse or a square wave to measure $h(t)$. Fig. 9-2 shows a real-time Fourier analyzer measuring $H(f)$.

An alternative description of $H(f)$ can be expressed in terms of the input and output signals,

$$H(f) = \frac{S_2(f)}{S_1(f)} \; .$$

Another interesting and sometimes useful description of $H(f)$ can be obtained in terms of the cross-power spectrum of the input and output, divided by the power spectrum of the input. To see this, simply multiply the equation above as follows,

$$H(f) = \frac{S_2(f)}{S_1(f)} \frac{S_1^*(f)}{S_1^*(f)} = \frac{G_{2,1}(f)}{G_{1,1}(f)} \; .$$

Fig. 9-2. A real-time Fourier analyzer can be used to measure an acoustic system function. Reverberation time is also easily measured by this method. See Chapter 4, Fig. 4-16, for direct measurement of impulse response. (Courtesy Brüel & Kjaer Instruments, Decatur, GA)

As usual the asterisk indicates the complex conjugate. This equation is often used in the process called *system identification*. We will discuss this in more detail in Chapter 10.

SYS-FUN, System Function Worksheet

Worksheet Organization

Row 4 is the Time axis. As it is set up here, there are 128 points running from zero to 127. For the FFT operation in spreadsheets it is necessary to have the data set in an integer power of 2. If you need more than 128 points then rotate the worksheet 90 degrees so that the data run in columns. You can get a maximum of 8192 cells in the vertical direction. If your data set is not an integer power of two then put zeros at the end of the set to pad it up to the next power of two. For example, if your data set has 100 points, add 28 zeros. Figure 9-3 shows the Home screen.

Row 6 is the Impulse Response $h(t)$. For this example the formula in Cell A6 is @EXP(-A4/100). This formula is copied to Cells A6..DX6. You can experiment with other impulse responses. The "@" symbol is not used in *Excel*. See Fig. 9-4.

Row 8 is the computed System Function, using the FFT. Note that the data in this row are complex. Put the cursor on cell B8, for example, to view the complex number. See Fig. 9-5.

Row 10 is the computed Impulse Response $h(t)$. The data in this row are obtained by using the IFFT on row 8. Note that the data in row 10 are complex.

Row 12 deletes the small errors due to the FFT and IFFT operations. The formula in Cell A12 is @IMREAL(A10). This is copied to Cells A12..DX12.

Row 14 contains the imaginary part of row 10. The formula in Cell A14 is @IMAGINARY(A10). This is copied to cells A14..DX14.

Row 16 computes the magnitude of the System Function $H(f)$. The formula in cell A16 is @IMABS(A8). This is copied to Cells A16..DX16.

Row 18 contains the phase of $H(f)$ in degrees. Cell A18 contains the formula @DEGREES(@IMARGUMENT(A8)), and this is copied to Cells A18..DX18.

Row 20 computes the error, input $h(t)$ – output $h(t)$. The formula in cell A20 is +A6 – A12. See Fig. 9-6.

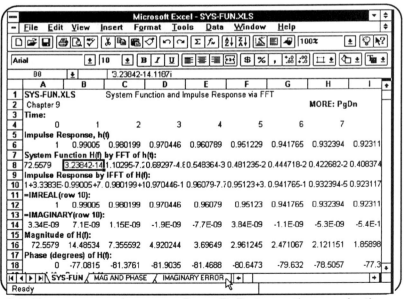

Fig. 9-3. Home screen of SYS-FUN. Press PgDn to see the rows for the centered System Function. The cursor on cell B8 shows its complex number in the formula editing slot below the icons. See your spreadsheet manual or on-line help for instructions on how to access and use the FFT.

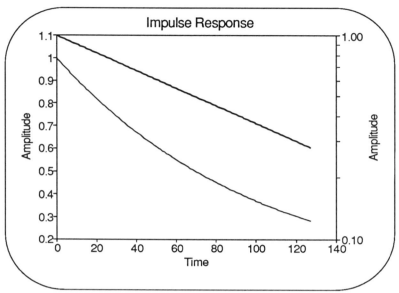

Fig. 9-4. A simple exponential decay is used as $h(t)$ in the example on the diskette. This is typical of a single-pole low-pass filter. Experiment with different time constants and see the effects on $H(f)$. Thin line: Left y-axis.

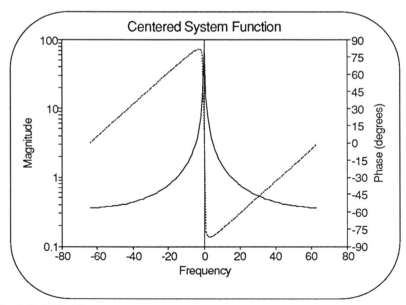

Fig. 9-5. This $H(f)$ corresponds to the $h(t)$ in Fig. 9-4. Observe how the spectrum expands as you decrease the time constant in $h(t)$, and vice versa.

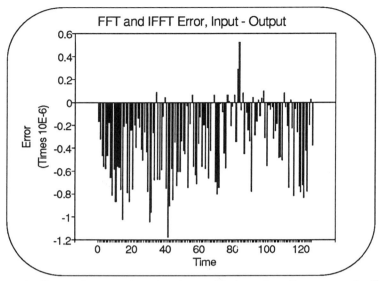

Fig. 9-6. Errors in transforming from the Impulse Response to the System Function, and then inverse-transforming back to the Impulse Response. Results of using the FFT and IFFT are certainly acceptable here.

Row 24 produces a checkerboarded $h(t)$ for viewing a centered $H(f)$. The formula in Cell A24 is A6*@COS(@PI*A4). This is copied to Cells A24..DX24.

Row 26 is the centered $H(f)$, obtained by using the FFT on row 24.

Row 28 contains the magnitude of the centered $H(f)$. The formula in Cell A28 is @IMABS(A26). This is copied to Cells A26..DX26.

Row 30 contains the phase of the centered $H(f)$. The formula in Cell A30 is @DEGREES(@IMARGUMENT(A26)), and this is copied to Cells A30..DX30.

Convolution

What is it?

Convolution is a process that appears everywhere. When a signal passes through or interacts with a system the result is usually a convolution. When you speak, your voice is the result of pulses from your vocal cords convolved with the resonances of your oral and nasal cavities. Convolution is the *fundamental linear filtering operation.*

In the original data domain convolution is a shift, multiply, and add process involving two functions. You can perform this operation in any domain, but it turns out that if the convolution occurs in the time domain then the operation can be performed more simply in the frequency domain. The converse is also true.

The reason for this lies in the *convolution theorem*, which says that convolution in one domain is simply multiplication in the Fourier transform domain. Conversely, if two functions are multiplied in one domain, this implies that they are convolved in the transform domain. Analytical operations with the Fourier transform can sometimes be mathematically challenging, but the FFT operation has brought the blessings of the Fourier transform to ordinary mortals.

In the frequency domain the spectrum of the output is simply the product of the System Function times the spectrum of the input,

$$S_{out}(f) \; = \; H(f)S_{in}(f) \; .$$

In general, all functions in the equation above are complex, so multiplication in the spreadsheet must use the IMPRODUCT function. This is the method that we will use in the CONVOLVE worksheet.

Convolution in the Time Domain

For small data sets it may be faster and simpler to perform the convolution in the original data domain, avoiding errors of the FFT. This method is used in the worksheets DIGIFILT and DIGIFIL2 for the FIR filters. This is particularly simple in a spreadsheet because of the ease with which the spreadsheet slides the time-reversed impulse response across the input signal,

$$S_{out}(t) \; = \; \frac{1}{K} \sum_{n=0}^{N-1} \; S_{in}(n) \; h(t-n) \; .$$

Here, K is the normalization factor. Some new spreadsheets use this method of convolution in data smoothing by computing a moving average. Look for this in the Analysis Tools of your spreadsheet.

CONVOLVE, Convolution Worksheet

This worksheet performs convolution using the FFT. This method is often called "circular convolution" for reasons we have seen in Chapter 8; the FFT imposes a periodicity on data that may not be periodic. Fig. 9-7 shows the Home screen of this worksheet.

Worksheet Organization

Cell C3 contains the variance of the input signal (see Appendix 2 for the meaning of variance). The formula in Cell C3 is @VAR(A14..DX14).

Cell F3 contains the variance of the output signal. The formula in Cell F3 is @VAR(A22..DX22). Note how the low-pass filter has reduced the variance.

Row 6 is the Time axis. As it is set up here, there are 128 points running from zero to 127. For the FFT operation in spreadsheets it is necessary to have the data set in an integer power of 2. See more about this in row 4 of SYS-FUN worksheet organization.

Row 8 is the Impulse Response $h(t)$. In this example the $h(t)$ is a simple exponential decay, typical of a single-pole low-pass filter.

Cell G7 is the normalization factor for $h(t)$. The formula is @SUM(A8..DX8), which simply adds all the points in $h(t)$.

Row 10 is the normalized $h(t)$. The formula in Cell A10 is +A8/$G$7. This formula is copied to Cells A10..DX10.

Row 12 is the System Function $H(f)$. The data in this row are obtained by using the FFT on row 10. Note that the data in row 12 are complex. Put the cursor on cell B12, for example, and see what the whole complex number looks like.

Row 14 is the input signal. You can input anything you want. At present the data in this row are zeros, except for 1's in AK14..BG14. This represents a sampled single pulse. Experiment with convolution; first try changing the length of the pulse and then use other types of signals.

Figures 9-8 and 9-9 show results for several pulses and a pulsed sine wave.

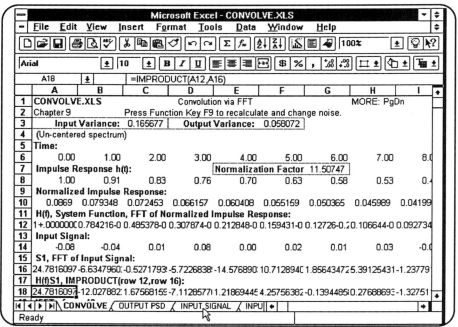

Fig. 9-7. Home screen of CONVOLVE . Formula in cell A18 shows how the output spectrum is computed. $h(t)$ and $H(f)$ are similar to those in Figs. 9-4 and 9-5.

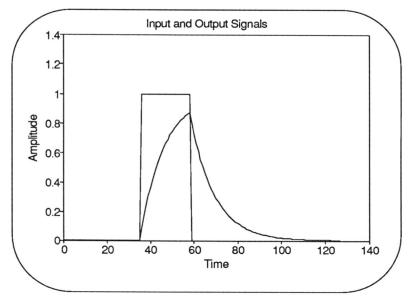

Fig. 9-8. The input rectangular pulse is transformed into an exponential rise and decay by the system, which behaves like a simple low-pass filter.

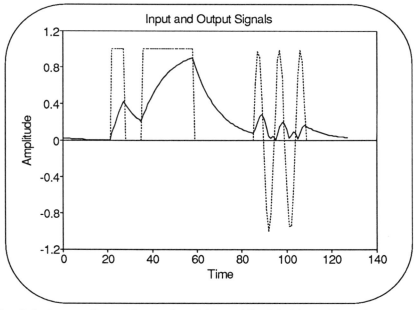

Fig. 9-9. A complicated input signal (dotted line) is blurred by a low-pass filter. High-frequency components in pulsed sine are greatly reduced.

Row 16 if the FFT of the input signal in row 14.

Row 18 multiplies the System Function $H(f)$ by the FFT of the input. The formula in Cell A18 is @IMPRODUCT(A12,A16). This formula is copied to Cells A18..DX18.

Row 20 is the IFFT of row 18. This gets you back to the original data domain. Observe that the data in this row are generally complex but the imaginary part is extremely small, compared with the real part.

Row 22 deletes the small errors due to the FFT and IFFT operations. The formula in Cell A22 is @IMREAL(A20). This is copied to Cells A22..DX22.

Row 25 computes the input power spectrum. The formula in Cell A25 is @IMABS(A16)^2. This is copied to Cells A25..DX25.

Row 27 is the output power spectrum. The formula in Cell A27 is @IMABS(A18)^2, and this is copied to Cells A27..DX27.

Row 29 is the Gain in dB. The formula in Cell A29 is 10*@LOG(A27/A25), and this is copied to Cells A29..DX29.

In Fig. 9-8 the input signal (a rectangular pulse) is convolved by an exponential-decay Impulse Response $h(t)$, resulting in a distorted (convolved) output signal. Observed that causality is preserved; there is no output before the input starts. However, there is some output after the input stops. If you use a longer decay time you will see the trailing edge of the "previous" pulse coming in from the left, because the FFT produces circular convolution, i.e., imposed periodicity.

Deconvolution

What is it?

Deconvolution is the "undo" operation for convolution. If an undesirable convolution occurs due to systematic distortion (amplitude and phase) it is often possible to restore the original to an acceptable condition. Checking engineering and scientific databases reveals that hundreds of papers are published each year on this subject. Deconvolution is used for a wide variety of problems. For example, it has been used to restore old phonograph recordings of Enrico Caruso and, in image processing, to restore photographs that were blurred due to camera motion or improper focus.

Audio recordings that are distorted due to reverberation can often be made more intelligible by "blind" deconvolution, in which only the magnitude of the system function is known or estimated. The success of this process is due to the fact that human hearing is not based on phase information (except for low-frequency direction-finding). Deconvolution has been used in theaters and concert halls, and in surveillance operations where a microwave or laser beam was reflected from a window in a building, phone booth, or automobile. A window vibrates due to the sound waves reflected from it, but a window also has many intrinsic natural resonances so it acts like a complicated filter.

Deconvolution, used extensively in astronomy (Hubble Space Telescope), radar, sonar, geology, and petroleum exploration, is simple in the frequency domain,

$$S_1(f) = \frac{S_2(f)}{H(f)} \ .$$

You measure the distorted output spectrum $S_2(f)$ and you measure or estimate the System Function $H(f)$ responsible for the distortion. If you don't know $H(f)$ you can round up the usual suspects and try them to see what sounds or looks best. Be careful! If there are zeros in $H(f)$ the deconvolution process can blow up. It is unlikely that the sampled version of $H(f)$ will hit a perfect zero, but it can happen.

A small amount of noise $N(f)$ can be added to $H(f)$ to reduce the probability of blow-up. This works best if $N(f)$ is the spectrum of the actual noise in the system,

$$S_1(f) = \frac{S_2(f)}{H(f) + N(f)} \ .$$

This has another advantage. At frequencies where the noise is small, ordinary deconvolution is carried out. Where the noise is large, $S_1(f)$ is minimized. Deconvolution can also be performed in the original data domain but it is usually more complicated (see Digging Deeper at the end of this chapter).

DECONVOL, Deconvolution Worksheet

This worksheet performs deconvolution using the FFT. Figure 9-10 shows the Home screen of this worksheet. Note that Cell H11 contains the computed normalization factor for the Impulse Response. Figures 9-11 and 9-12 show the results.

Worksheet Organization

Row 4 is the Time axis. As it is set up here, there are 128 points running from zero to 127. For the FFT operation in spreadsheets it is necessary to have the data set in an integer power of 2. If you need more than 128 points then rotate the worksheet 90 degrees so that the data run in columns. You can get a maximum of 8192 cells in the vertical direction. If your data set is not an integer power of two then put zeros at the end of the set to pad it up to the next power of two. For example, if your data set has 100 points, add 28 zeros.

Row 6 is the output waveform $s_2(t)$. In this example row 6 has been generated internally (see rows 33-39). The formula in cell A6 is IMREAL(A39), and this is copied to cells A6..DX6.

Row 8 is the output spectrum, obtained by the FFT of row 6.

Row 10 is the Impulse Response $h(t)$, which must be known from measurement or design, or estimated. If you don't know $h(t)$ it is common practice to try several to see which gives the best result.

Cell H11 is the normalization factor. The formula is SUM(A10..DX10).

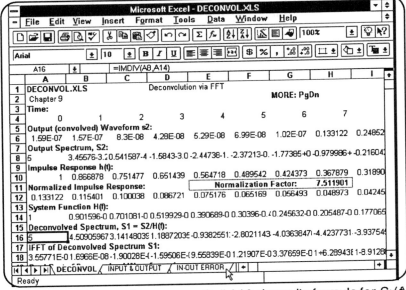

Fig. 9-10. Home screen of DECONVOL. Cell A16 shows its formula for $S_1(f)$ in the line above the spreadsheet.

Row 12 is the normalized Impulse Response. For an exponential decay the formula in Cell A12 is +A10/$H$11. This formula is copied to Cells A12..DX12. Experiment with deconvolution; first try changing the time constant in $h(t)$, and then use other types of signals.

Row 14 is the System Function $H(f)$, which is obtained by the FFT of row 12.

Row 16 divides the output spectrum by the System Function $H(f)$. The formula in Cell A16 is IMDIV(A8,A14). This formula is copied to Cells A16..DX16.

Row 18 is the deconvolved waveform, computed by the IFFT of row 16. This gets you back to the original data domain. Observe that the data in this row are generally complex but the imaginary part is extremely small, compared with the real part.

Row 20 deletes the small errors due to the FFT and IFFT operations. The formula in Cell A20 is IMREAL(A18). This is copied to Cells A20..DX20.

Row 22 computes the deconvolution error, original signal minus deconvolved signal. The formula in Cell A22 is +A35 – A20. This is copied to Cells A22..DX22.

Rows 25 - 29 are used to obtain vertical displacement for graphing.

Rows 33 - 39 generate an internal test signal, a pulse convolved with $h(t)$.

Row 35 contains the original test input. On the diskette this is a pulse; try different signals, and add some random noise.

Row 37 is the spectrum of row 35.

Row 39 is the IFFT of the convolved spectrum (IMPRODUCT of row 37 and row 14) to obtain the test output waveform. Real part of row 39 is used in row 6.

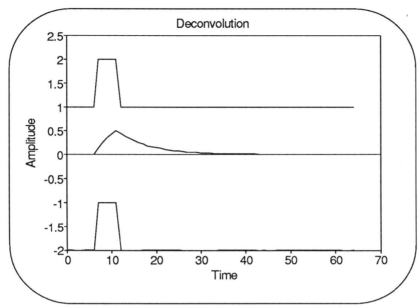

Fig. 9-11. Top trace: Original signal. Middle trace: Convolved signal. Bottom trace: Deconvolved signal. Try experimenting with other signals and $h(t)$'s.

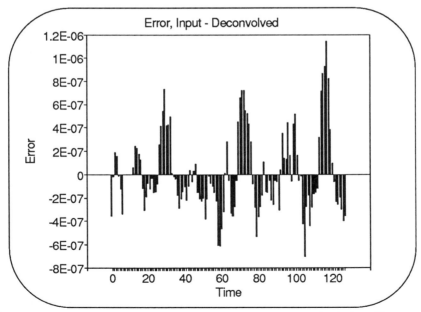

Fig. 9-12. Errors in this deconvolution example. The original signal is recovered with excellent accuracy.

Data Windows

What Are they?
Data windows are functions that are usually larger in the middle than at the edges.
A rectangular (or uniform) window is constant from edge to edge, and goes to zero
at the edges. We have seen that the spectrum of a rectangular pulse is spread out
over a wide band of frequencies, so it follows that starting and stopping a waveform
will spread its natural spectrum.

In other words, observing a signal for a limited time produces spectrum leakage
into frequencies not contained in the original signal. In order to reduce this leakage
and enhance spectrum resolution a variety of windows have been invented. There is
no "best" window, but the rectangular window is probably the worst. Each window
has its own advantages and disadvantages, and there are many trade-offs to be
considered. Window figures of merit have been summarized by Harris (see Digging
Deeper section at the end of this chapter). It is useful to examine a waveform through
a variety of windows to see what you can gain and what you have to lose. A window
function multiplies a signal, so this implies that the window and signal are convolved
in the frequency domain.

WINDOWME, Data Window Worksheet
The Home screen of the WINDOWME worksheet is shown Fig. 9-13 .

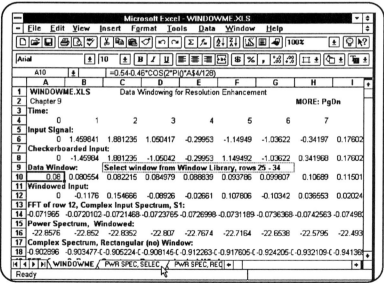

Fig. 9-13. Home screen of WINDOWME. When you view this on your monitor,
press PgDn to see the window library. You can easily add your favorite windows
to the portfolio.

Worksheet Organization

Row 4 is the Time axis. As it is set up here, there are 128 points running from zero to 127. For the FFT operation in spreadsheets it is necessary to have the data set in an integer power of 2. If you need more than 128 points then rotate the worksheet 90 degrees so that the data run in columns. You can get a maximum of 8192 cells in the vertical direction. If your data set is not an integer power of two then put zeros at the end of the set to pad it up to the next power of two. For example, if your data set has 100 points, add 28 zeros. This message has been repeated in case you didn't read it before.

Row 6 is the input waveform. You can use any waveform in this row. At present, the diskette has an internally-generated waveform for test purposes. This waveform is the sum of two sine waves of the same amplitude but different frequency. This two-tone test signal is commonly used in checking for intermodulation distortion (caused by system nonlinearity). The formula for the two-tone signal in Cell A6 is @SIN(A4)+@SIN(A4/1.5), and this is copied to Cells A6..DX6. Figure 9-14 shows the original signal in this worksheet; Fig. 9-15 shows the windowed signal.

Row 8 is the checkerboarded input, used to center the spectrum. The formula in Cell A8 is +A6*@COS(A4*@PI). This is copied to Cells A8..DX8.

Row 10 is the data window selected from the Window Library, rows 26-34. For example, to select the Hamming window copy Cell A30 to Cells A10..DX10.

Row 12 is the checkerboarded, windowed input waveform. The formula in Cell A12 is +A8*A10. This is copied to Cells A12..DX12.

Row 14 if the FFT of the signal in row 12. This produces the centered complex spectrum of the windowed input signal.

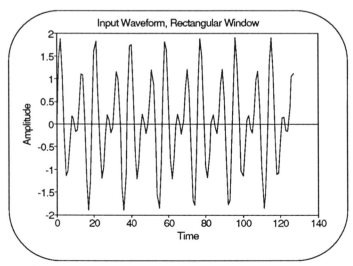

Fig. 9-14. Original signal, consisting of two sine waves of slightly different frequency. Using no window is equivalent to using a rectangular window.

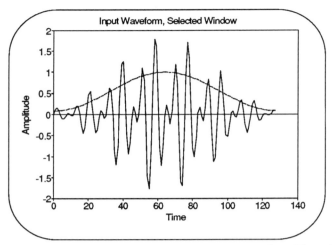

Fig. 9-15. Original signal seen through a Hamming window. The window (dotted line) has diminished the importance of data near the start and stop.

Row 16 computes the power spectrum of row 12. The formula in A16 is 20*@LOG(@IMABS(A14)), and this formula is copied to Cells A16..DX16.

Row 18 is the FFT of row 8. The computes the complex spectrum of the checkerboarded input signal with a rectangular (uniform) window.

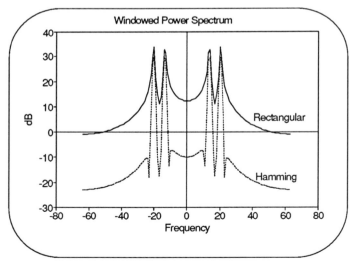

Fig. 9-16. Power spectra comparison of the signal with a rectangular window (solid line) and a Hamming window (dotted line). The Hamming window has dramatically reduced spectrum leakage, enhancing the frequency resolution. This price that is paid is a few dB reduction in peak power.

Row 20 computes the power spectrum of row 18 (see Fig. 9-16). The formula in A20 is 20*@LOG(@IMABS(A18)), and this is copied to Cells A20..DX20.

Row 24 is the input with the selected window, without checkerboarding. The formula in Cell A24 is A6*A10, and this is copied to Cells A24..DX24.

Cell 28 contains the von Hann data window. The formula in Cell A28 is 0.5–0.5*@COS(2*@PI*A$4/128).

Cell 30 contains the Hamming data window. The formula in Cell A30 is 0.54–0.46*@COS(2*@PI*A$4/128).

Cell 32 contains the Blackman-Harris window. The formula in Cell A32 is 0.358–0.488*@COS(2*@PI*A$4/128)+0.141*@COS(4*@PI*A$4/128)–0.011 *@COS(6*@PI*A$4/128).

Cell 34 contains the Kaiser-Bessel data window. The formula in Cell A34 is 0.402–0.498*@COS(2*@PI*A$4/128)–0.098*@COS(4*@PI*A$4/128)+0.001 *@COS(6*@PI*A$4/128).

FREQ-RES, Frequency Resolution Worksheet

This worksheet lets you experiment with the frequency resolution of the FFT process. You will discover that there is a lower limit to the ability to distinguish two sinusoidal signals that are closely spaced in frequency. A major reason for this is the start, stop, and finite duration of the measurement process, i.e., the length of the data set and the abrupt onset and termination of data acquisition. This type of data acquisition is equivalent to the use of a rectangular window, and this results in spreading or leakage of energy into nearby frequencies. Now that we have some data windows we can use them to test frequency resolution and compare the trade-offs involved in enhancing resolution. In this worksheet we will use two sine waves, very close in frequency, but one sine wave will have half the amplitude of the other. We will compare a rectangular window with the Blackman-Harris window, and we will see the remarkable advantages of the latter.

Worksheet Organization

The Home screen is shown in Fig. 9-17. Press PgDn to see the second screen. You can select other windows and compare them with each other. Row 4 is the Time axis. As it is set up here, there are 128 points running from zero to 127. For the FFT operation in spreadsheets the data set must be an integer power of 2.

If you need more than 128 points then rotate the worksheet 90 degrees so that the data run in columns. A maximum of 8192 cells are in the vertical direction. If your data set is not an integer power of two then put zeros at the end of the set to pad it up to the next power of two. For example, if your data set has 100 points, add 28 zeros. This message has been repeated in case you didn't read the previous instructions. Adding zeros in one domain produces a smoother data set in the transform domain, although it is obvious that no information is added.

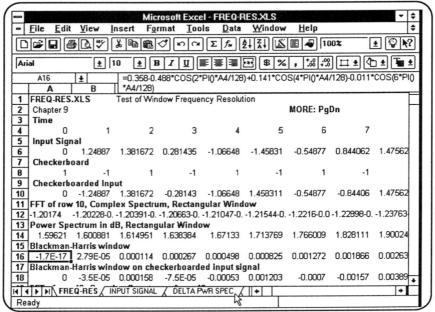

Fig. 9-17. Home screen of FREQ-RES. The cursor on cell A16 displays the Blackman-Harris formula.

Row 6 is the input waveform. As supplied on the diskette the waveform is two sine waves, close in frequency but of different amplitudes. The formula in Cell A6 is @SIN(A4)+0.5*@SIN(A4/1.5). This is copied to Cells A6..DX6. See Fig. 9-18.

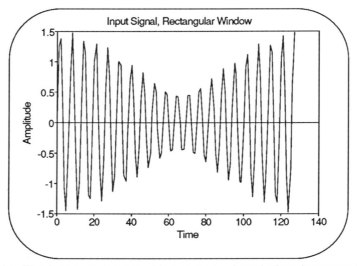

Fig. 9-18. The input waveform with a rectangular window (see row 6).

Row 8 is the checkerboard to center the spectrum. The formula in Cell A8 is @COS(A4*@PI). This formula is copied to Cells A8..DX8.

Row 10 is the checkerboarded input waveform with the rectangular (uniform) window. The formula in Cell A10 is +A6*A8, and this is copied to Cells A10..DX10.

Row 12 is the spectrum of the checkerboarded input with the rectangular window. This row is the FFT of row 10. See Fig. 9-19.

Row 14 is the power spectrum using the rectangular window. The formula in Cell A14 is @IMABS(A12)^2, and this is copied to Cells A14..DX14.

Row 16 contains the Blackman-Harris window. You can see its formula in the formula-editing row of Fig. 9-17, just below the bottom row of icons. This formula is copied to Cells A16..DX16.

Row 18 is the checkerboarded input waveform multiplied by the Blackman-Harris window. The formula in Cell A18 is +A10*A16.

Row 20 is the centered complex spectrum of the input waveform modified by the Blackman-Harris window. This is the FFT of row 18.

Row 22 contains the power spectrum of row 20. The formula in Cell A22 is @IMABS(A20)^2. This is copied to Cells A22..DX22. See Figs. 9-20 and 9-21.

Row 24 is the difference in the power spectra of the rectangular and Blackman-Harris windows. The formula in Cell A24 is +A14 − A22, and this is copied to Cells A24..DX24. See Fig. 9-22.

Row 28 is the Blackman-Harris window on the un-checkerboarded input waveform. This is for graphical purposes. The formula in Cell A28 is +A16*A6, and this is copied to Cells A28..DX28.

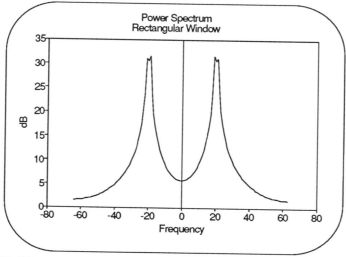

Fig. 9-19. The centered spectrum of two closely-spaced sine waves. One sine wave has half the amplitude of the other. The signals are barely resolved.

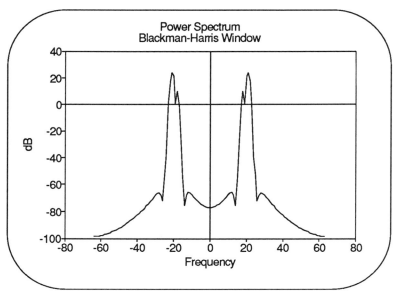

Fig. 9-20. The same signals seen through a Blackman-Harris window. The smaller, lower-frequency signal is well-resolved and clearly has the smaller amplitude. In Fig. 9-19 the uniform window shows the amplitudes incorrectly.

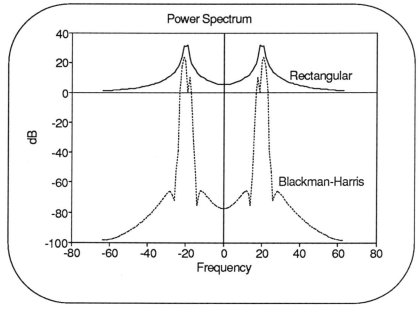

Fig. 9-21. The power spectra of Figs. 9-19 and 9-20 are plotted here on the same scale. No vertical displacement is imposed; these are the actual spectra.

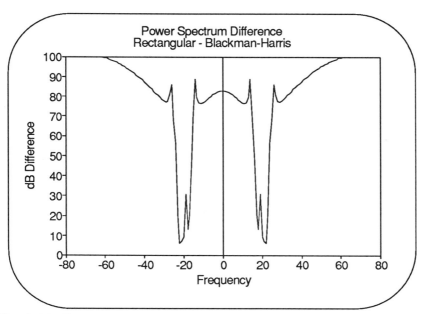

Fig. 9-22. The spectrum difference, rectangular minus Blackman-Harris, is a dramatic illustration of resolution improvement due to windowing. There is about 80 dB improvement close to the central frequency and the smaller signal is clearly present. This is the graph of row 24. Graph name: *Delta Spectrum*.

Digging Deeper

Digital Signal Processing, A. V. Oppenheim and R. W. Schafer (Prentice-Hall, Englewood Cliffs, New Jersey, 1975) ISBN 0-13-214635-5. This is an in-depth introduction to the mathematical foundations of digital signal processing, at the senior and graduate level.

Advanced Topics in Signal Processing, J. S. Lim and A. V. Oppenheim, Eds. (Prentice-Hall, Englewood Cliffs, New Jersey, 1988) ISBN 0-13-013129-6

Handbook of Digital Signal Processing, D. F. Elliott, Ed. (Academic Press, San Diego, California 1987) ISBN 0-12-237075-9.

SSP: The Spreadsheet Signal Processor, S. C. Bloch (Prentice-Hall, Englewood Cliffs, New Jersey, 1992) ISBN 0-13-830506-4. Chapter 2 has many examples of convolution in the original data domain (without using the FFT). Chapter 4 discusses data windows, and Chapter 11 has examples of deconvolution in the original data domain (without using the FFT). Two diskettes are included.

Deconvolution, P. A. Jansson (Academic Press, Orlando, Florida, 1984) ISBN 0-12-380220-2. The main emphasis of this book is on spectroscopy, but the methods are universal.

F. J. Harris, "On the use of windows for harmonic analysis with the discrete Fourier transform," *Proceedings of the IEEE*, **66**, pp. 51 - 83 (1978). This paper contains a plethora of information on 23 types of data windows.

T. Holman, "Motion-picture theater sound system performance: New studies of the B-chain," *Society of Motion Picture and Television Engineers Journal*, **103**, pp. 136 - 149 (1994). This provides a good illustration of the use of the inverse FFT to obtain the transient response of a system, which includes reflections in this case.

The Society of Motion Picture and Television Engineers (SMPTE) was founded in 1916. It publishes the *SMPTE Journal* and is a good source for current Standards in entertainment and telecommunications. The SMPTE hosts several conferences every year where the latest audio and image processing hardware and software are displayed. Society of Motion Picture and Television Engineers, Inc., 595 W. Hartsdale Avenue, White Plains, New York 10607-1824. Phone (914) 761-1100; FAX (914) 761-3115.

Caruso, A Legendary Performer (RCA Records, New York,) Catalog Number CRM1-1749. This is a collection of original recordings of Enrico Caruso that were digitally deconvolved and restored by Dr. Thomas G. Stockham, Jr.

Chapter 10

Advanced FFT Operations, II

What This Chapter is About

In Chapter 8 we explored a few basic operations needed to transform in and out of the frequency domain using the FFT and IFFT. In Chapter 9 we showed how to perform some FFT operations involving a single-channel input. With that preparation we are ready to plunge more deeply into the frequency domain and learn some useful functions and operations involving multi-channel inputs. We will also demonstrate a spreadsheet trick with the @IF function that will enable you to obtain digital readouts of time delays and signal strengths in the presence of random noise and other interfering signals.

All of the worksheets in this chapter use the FFT operation. Even if your spreadsheet does not have the FFT you can observe how the functions operate, but you will not be able to make changes in the worksheet. FFT add-ins are available for *Lotus 1-2-3* and *Excel*. *Excel*'s version 4 and newer, and *Quattro Pro* 5.0 and newer, have a built-in FFT. In this Chapter we will describe worksheets so you can experiment with:

- Cross-spectrum function

- Cross-correlation and auto-correlation functions

- Coherence (MSC) function

- SNR function

- System identification

- Cepstrum function

- Analytic and nonanalytic signals

- Phased arrays

These worksheets use some of the complex @-functions described in Chapter 8. If your spreadsheet doesn't have these functions you can still see how they work but, of course, you can't make changes. The @-symbol is not used in *Excel*. The XLS worksheets on the diskette are ready to run in *Excel*; for *Quattro Pro* import the WK1 files and modify formulas as shown in the Worksheet Organization. You may also want to modify graphs to take advantage of the right-hand *y*-axis and log scaling capabilities of *Quattro Pro* which are not available in the WK1 format.

Cross-Spectrum

What Is It?

The cross-spectrum is the product of two complex spectra, so it is generally a complex function of frequency. The cross-spectrum is interesting and useful in itself, but it is also the well-known secret ingredient used in other things such as the cross-correlation functions, coherence function, and system identification. For many advanced operations the cross-spectrum is a building block, often the first step. The cross-power spectrum $G_{1,2}(f)$ is defined as,

$$G_{1,2}(f) = S_1^*(f)\, S_2(f) \, .$$

where, $S_1^*(f)$ is the complex conjugate of the spectrum of signal 1 and $S_2^*(f)$ is the spectrum of signal 2. We will use this form in the cross-correlation function. Sometimes we also need a cross-spectrum in which no complex conjugate is used. This type of cross-spectrum preserves phase information as in,

$$G_{1,2}(f) = S_1(f)\, S_2(f) \, .$$

What Does It Tell You?

The cross-spectrum contains only those frequencies that two signals have in common. For example, a low-pass filter decreases the high-frequency content of a signal and introduces phase distortion of the passed frequencies. The cross-spectrum of the filter's input and output shows how the input spectrum, whatever it is, becomes modified by the filter. This is useful if the input consists of a desired signal accompanied by random noise because the cross-spectrum will provide clues about how to enhance the signal by minimizing the noise, provided the spectral overlap is not excessive.

We have already used the second type of cross-spectrum in the CONVOLVE worksheet in Chapter 9. In that worksheet we needed the cross-spectrum $H(f)S_1(f)$, which we then inverse-transformed to get the output waveform. In other words, this cross-spectrum is actually the spectrum of the output; it contains only those frequencies common to both the input and the system function.

CROSSPEC, Cross-Spectrum Worksheet

Worksheet Organization

Row 4 is the Time axis. The number of points must be in the 2^n form. See the worksheet organization information in Chapter 9 for other details about this row.

Row 6 contains signal #1. As shown in Fig. 10-1, there is a rectangular pulse with amplitude 1 and duration 7 time units.

Row 8 contains signal #2. The worksheet on the diskette contains a signal similar to row 6, but starting much later.

Row 10 is the complex spectrum of signal #1.

Rpw 12 is the complex spectrum of signal #2.

Row 14 is the cross-power spectrum using the complex conjugate. The formula in Cell A14 is @IMPRODUCT(@IMCONJUGATE(A10),A12). This formula is copied to Cells A14..DX14.

Row 16 is the magnitude of row 14. The formula in Cell A16 is @IMABS(A14). This formula is copied to Cells A16..DX16.

Row 18 is the cross-power spectrum not using the complex conjugate. The formula in Cell A18 is @IMPRODUCT(A10,A12). This formula is copied to Cells A18..DX18.

The Home screen of CROSSPEC is shown in Fig. 10-1. A cross-spectrum example is shown in Figs. 10-2 and 10-3.

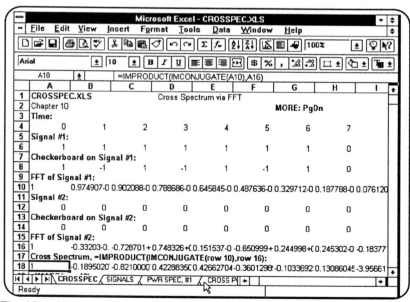

Fig. 10-1. Home screen of CROSSPEC. The cursor on cell A18 displays its formula in the cell-editing space below the icons.

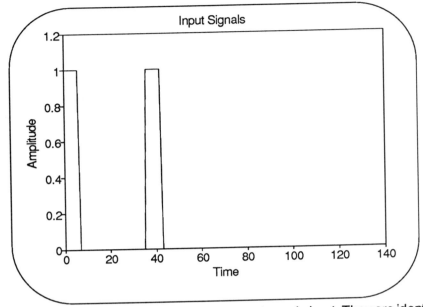

Fig. 10-2. These are the two pulses in the sample worksheet. They are identical except that they are displaced in time.

Row 20 is the magnitude of row 18. The formula in Cell A20 is @IMABS(A18). This formula is copied to Cells A20..DX20.

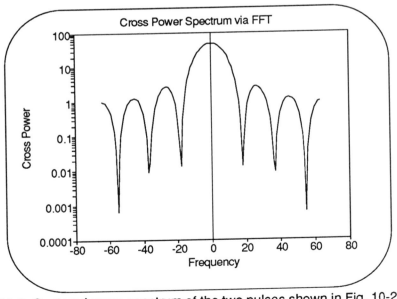

Fig. 10-3. Centered cross-spectrum of the two pulses shown in Fig. 10-2.

Cross-Correlation

What Is It?

The cross-correlation function $R_{1,2}(\tau)$ is a function of the time delay τ. The cross-power spectrum $G_{1,2}(f) = S_1^*(f)\, S_2(f)$ and the cross-correlation function $R_{1,2}(\tau)$ comprise a Fourier transform pair,

$$R_{1,2}(\tau) = \text{IFT}[G_{1,2}(f)]$$

$$G_{1,2}(f) = \text{FT}[R_{1,2}(\tau)]$$

where IFT is the inverse Fourier transform and FT is the forward Fourier transform. This important result is the Wiener-Khintchine theorem, a remarkable and useful relation. When you have the cross-power spectrum (using the complex conjugate) it is quick and easy to compute the cross-correlation function with the IFT. Of course, in the worksheets and in most practical computations the IFFT and FFT are used.

What Does It Tell You?

The cross-correlation function is useful for discovering hidden similarities in data and for measuring time delays. For example, if you know what signal you are looking for you can use the cross-correlation function to pick it out of noise and interfering signals. The auto-correlation function can discover hidden periodicities in data.

The cross-correlation function achieves its maximum value at a time delay that corresponds to the time of arrival of a signal that resembles a stored replica. The value of the cross-correlation function, at its maximum, is a measure of the similarity of the received signal to the stored replica.

In systems with multiple inputs and outputs the cross-correlation function lets you check out each input with each output to see which are most related. This is a simple wave of detecting cross-talk between different inputs. Signals buried deep in noise can be detected with amazing results. In image processing the cross-correlation function is used in matching smudged fingerprints and smashed bullets, among other things.

Accurate measurement of time delay in the presence of noise and interference is where the cross-correlation function really shines. Time delays can be determined even when measurements directly on waveforms are impossible. Specialized analog instruments have been invented to perform cross-correlation. Although they appear to be quite different, the lock-in amplifier and the box-car integrator are basically cross-correlators.

There is a close connection between cross-correlators and matched filters. Both are used in radar, sonar, and the Global Positioning System. The main difference is that the cross-correllator output is in time delay, whereas the matched filter output is in the same data domain as the input, whatever it is.

Auto-Correlation Function

The auto-correlation function is a real function of time delay τ. It provides an alternative method of computing power spectra. This function is like cross-correlation, but it refers to only one signal,

$$R_{1,1}(\tau) = \text{IFT}[G_{1,1}(f)]$$

$$G_{1,1}(f) = \text{FT}[R_{1,1}(\tau)] \ .$$

Correlation in the Time Domain

If your spreadsheet does not have the FFT, or if you would like to get away from the circular correlation problems, you can compute the correlation function directly in the original data domain in your spreadsheet. The auto-correlation function is defined as,

$$R_{1,1}(\tau) = \frac{1}{K} \sum_{n=0}^{N-1} s(n)\, s(n+\tau) \ .$$

The normalization factor $1/K$ is usually chosen as $1/N$. You can adjust it to produce a maximum value of one for $R_{1,1}(\tau)$, for a noiseless signal. If the two signals are different $R_{1,2}(\tau)$ is the cross-correlation function.

Correlation in the original data domain is very accurate, avoiding the circularity of the FFT. This shift, multiply, and add process can be fast for small data sets but it becomes unwieldy for long signals. An example is shown in CROSSCOR, row 37. See Fig. 10-9 for results.

CROSSCOR, Correlation Function Worksheet

Worksheet Organization

The Home screen of this worksheet is shown in Fig. 10-4.

Cell D3 displays the time delay of the signal, measured by the maximum value of the crosss-correlation function. The formula in this Cell is @MAX(A21..DX21). This Cell operates explicitly with row 21 and implicitly with row 5 (see row 21).

Cell G3 detects the maximum value of the normalized cross-correlation function. The formula in this Cell is @MAX(A19..DX19).

Row 5 is the Time axis. On the diskette it has a length of 128.

Row 7 is the stored replica of the expected signal.

Row 9 contains the input signal. On the diskette row 9 has a signal similar to the stored replica and random noise. You can adjust the noise and the location of the signal.

Row 11 is the FFT of the stored replica, row 7.

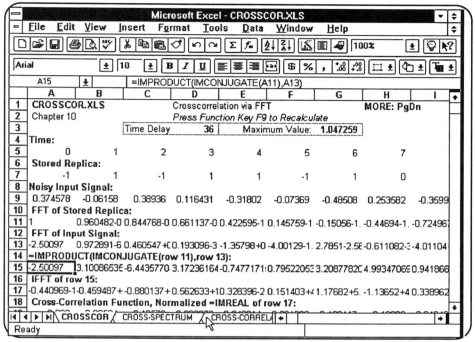

Fig 10-4. Home screen of CROSSCOR. Cell D3 shows the computed time of arrival of the expected signal, and cell G3 shows the computed peak value of the normalized cross-correlation function. The value in G3 is more than 1 because of a noise burst in the received signal. This is unusual; mostly this value is 1 or less.

Row 13 is the FFT of the input signal, row 9.

Row 15 computes the conjugated cross-spectrum. The formula in Cell A15 is @IMPRODUCT(@IMCONJUGATE(A11),A13). This formula is copied to Cells A15..DX15.

Row 17 is the IFFT of row 15.

Row 19 is the normalized cross-correlation function. In the diskette file the normalization factor is 7 because the replica has an amplitude of 1 and a length of 7. The formula in Cell A19 is @IMREAL(A17)/7 and this is copied to Cells A19..DX19.

Row 21 is the peak detector. The formula in Cell A21 is @IF(A19=$G$3,A5,0). This is copied to Cells A21..DX21. The @IF function works like this: If the numerical value in the Cell is equal to the value in Cell G3 (the maximum value of row 19) then @IF writes the numerical value of the associated Cell in row 5 (the time). Otherwise, @IF writes 0.

Figure 10-5 shows a result of cross-correlation of a noisy coded pulse group.

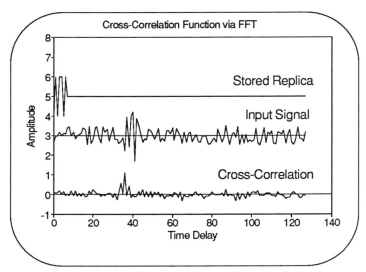

Fig. 10-5. Cross-correlation function for a coded pulse group in the presence of noise. The arrival time of the pulse group cannot be accurately determined by measuring the waveform but the sharp peak of the cross-correlation clearly marks the arrival of the signal.

Row 26 is used for graphing. The formula in Cell A26 is +A7+5. This is copied to Cells A26..DX26.

Row 28 is used for graphing. The formula in Cell A28 is +A9+3. This is copied to Cells A28..DX28.

Row 30 is used for graphing. Cell A30 contains 3, and this is copied to Cells A30..DX30.

Row 32 computes the magnitude of row 15, the cross-spectrum. The formula in Cell A32 is @IMABS(A15). This is copied to Cells A32..DX32.

Row 34 computes the phase of row 15. The formula in Cell A34 is @DE-GREES(@IMARGUMENT(A15)). This is copied to Cells A34..DX34.

Row 37 computes the normalized cross-correlation function without the FFT. You can see how simple and compact this is for a small signal. The formula in Cell A37 is (+$A7*A9+$B7*B9+$C7*C9+$D7*D9+$E7*E9+$F7*F9+$G7*G9)/7. This is copied to Cells A37..DX37. In row 37 the stored replica is dragged across the input signal; the normalized overlap product is computed at each time increment.

Row 39 is a peak detector similar to row 21. Time of arrival is detected in row 37.

Cell A41 contains the maximum value of the cross-correlation function, similar to Cell G3.

Cell A43 contains the time delay of the input signal, similar to Cell D3.

Figures 10-6 and 10-7 show cross-correlation of a noiseless pulsed sine wave.

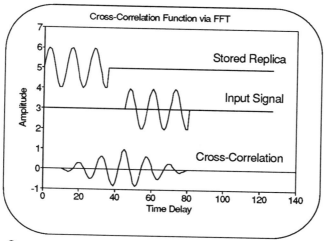

Fig. 10-6. Cross-correlation of a noiseless, pulsed sine wave. The maximum of the cross-correlation function occurs at the time of arrival of the input signal. If noise is present the time of arrival would be impossible to measure on the waveform itself because the signal starts at zero and rises slowly in noise.

Press the Recalculate Function Key (usually F9) to change the noise. You must perform the two FFTs and the IFFT again because they do not recalculate automatically. You may want to write a macro do this task automatically in your spreadsheet. Figures 10-8 and 10-9 show the results of time-domain (non-FFT) computation.

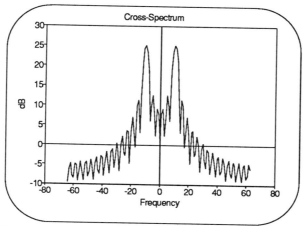

Fig. 10-7. The centered cross-spectrum of the pulsed sine wave shown in Fig. 10-6. Note that the center frequency has been shifted away from zero, to the frequency of the sine wave, and there are now two peaks, corresponding to positive and negative frequencies. Compare with the ordinary pulse which is centered at zero frequency, Figs. 8-20 and 10-3.

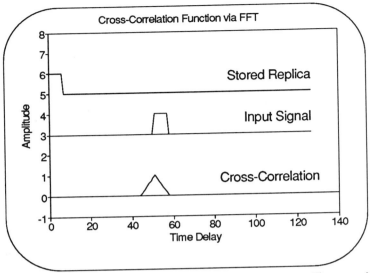

Fig. 10-8. Cross-correlation function of a noiseless pulse. The spectrum of this signal is centered at zero frequency. Note that the *envelope* of the cross-correlation function in Fig. 10-6 is a triangle with a peak when the signal arrives.

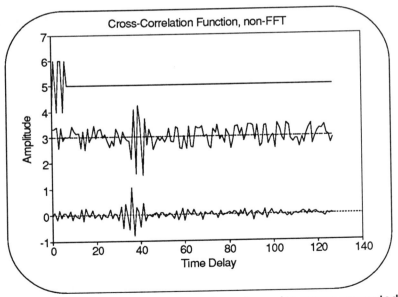

Fig. 10-9. Cross-correlation function of a noisy pulse group computed without the FFT. Compare with Fig. 10-5 which has the same signal but a different noise sample. Upper: Stored replica. Middle: Noisy signal. Lower: Cross-correlation sharp maximum shows accurate time delay of signal, even with noise.

Coherence (MSC) Function

What Is It?

The coherence function $\gamma^2(f)$ is a dimensionless real function of frequency composed of a cross-power spectrum and the two power spectra associated with the signals in the cross-power spectrum. It's value ranges from 0 to 1. The coherence function is usually defined as,

$$\gamma^2(f) = \frac{|S_{1,2}(f)|^2}{S_{1,1}(f)\, S_{2,2}(f)}$$

Some analysts call this the MSC, or magnitude-squared coherence function. Others simply call this the coherence function. The MSC can be used for itself but more often it is used in computing the SNR function, discussed below.

Because of round-off error and the unpredictability of samples of random noise the MSC may be slightly greater than one for a particluar noise sample. If you encounter this try another noise sample. Press function key F9 to recalculate with new noise.

What Does It Tell You?

The MSC function is important because at *each frequency* it represents the fraction of the system output power directly related to a particular input power. A value of 1 indicates that all of the output is due to the input. A value less than 1 can be caused by one or more factors:

- Noise may contaminate the measurements

- Other inputs may be present

- The system may have nonlinearity

The possibility of detecting the presence of inputs other than your own is a valuable test procedure for cross-coupling.

SNR Function

What Is It?

The SNR function is a simple extension of the MSC function $\gamma^2(f)$, and is defined as,

$$SNR(f) = \frac{\gamma^2(f)}{1 - \gamma^2(f)} \; .$$

What Does It Tell You?

The SNR function gives a quantitative measure of the signal/noise ratio at each frequency. (Recall that the ordinary SNR is not a function; it is number that is the ratio of the time-averaged signal power to the time-averaged noise power.)

The SNR function is often expressed in dB. Be careful of that denominator! It can blow up for highly coherent signals. Also, for some unusual random noise samples the MSC may be greater than one. Press function key F9 and try again with another sample. This is usually not a problem with signals acquired from the real world because things out there are pretty messy.

SNR-MSC, SNR and Coherence Worksheet

This worksheet uses two FFT operations. Remember to perform the FFT every time the signal or noise changes, because they do not automatically recalculate. The Home screen is shown in Fig. 10-10.

Worksheet Organization

Cell B3 shows the maximum value of the SNR function. The formula in this cell is @MAX(A40..DX40).

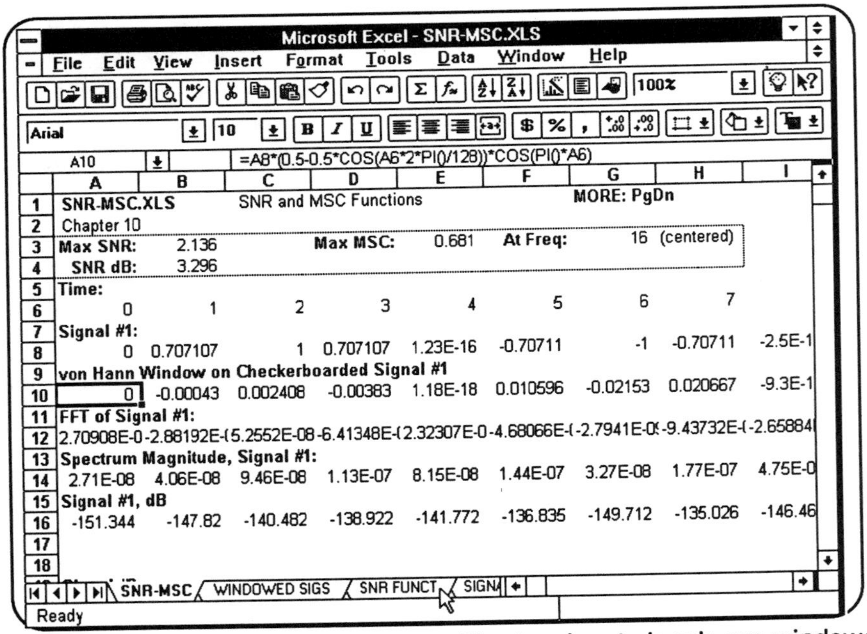

Fig. 10-10. Home screen of SNR-MSC. The two input signals are windowed to enhance frequency resolution and checkerboarded to center the spectra, SNR function and MSC function. Cells B3, E3, G3, and B4 hold computed information. Note: "CALC" at the bottom shows when automatic recalculation is off.

Cell E3 shows the maximum value of the MSC function. The formula in this Cell is @MAX(A38..DX38).

Cell G3 displays the (positive) frequency at which the SNR and MSC achieve their maximum values. The frequency here goes from 0 to 128.

Cell B4 shows the value of the SNR function in dB. The formula in this Cell is @MAX(A42..DX42).

Row 6 is the Time axis. In the diskette the axis is 128 points, from A6 to DX6.

Row 8 contains signal #1, a sine wave. The formula contained in Cell A8 is @SIN(@*@PI*A6/8). This is copied to Cells A8..DX8. See Fig. 10-11.

Row 10 performs a dual job. It imposes a von Hann window on signal #1 to enahnce frequency resolution, and it checkerboards the signal to center the spectrum. The formula in Cell A10 is:

+A8*(0.5–0.5*@COS(A6*2*@PI/128)*@COS(@PI*A6).) This is copied to Cells A10..DX10.

Row 12 is the spectrum of windowed, checkerboarded signal #1, the FFT of row 10.

Row 14 contains the un-normalized spectrum magnitude of signal #1. The formula in Cell A14 is @IMABS(A12). This is copied to Cells A14..DX14.

Row 16 contains the spectrum of signal #1 in dB. The formula in Cell A16 is 20*@LOG(A14). This is copied to Cells A16..DX16.

Row 20 contains signal #2, a noisy sine wave. The formula in Cell A20 is @SIN(@*@PI*A6/8) +@RAND*6-3. This is copied to Cells A20..DX20.

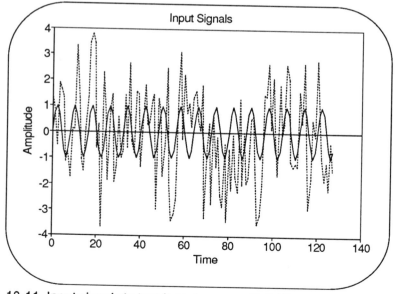

Fig. 10-11. Input signals to test for coherence. Solid line, signal #1. Dotted line, signal #2. Overall, there appears to be little coherence between the signals.

Row 22 is similar to row 10. It imposes a von Hann window on signal #2 to enahnce frequency resolution, and it checkerboards the signal to center the spectrum. The formula in Cell A22 is:

+A20*(0.5-0.5*@COS(A6*2*@PI/128)*@COS(@PI*A6).) This is copied to Cells A22..DX22.

Row 24 is the spectrum of windowed, checkerboarded signal #2, the FFT of row 22.

Row 26 contains the un-normalized spectrum magnitude of signal #2. The formula in Cell A26 is @IMABS(A24). This is copied to Cells A26..DX26.

Row 28 contains the spectrum of signal #2 in dB. The formula in Cell A28 is 20*@LOG(A26). This is copied to Cells A28..DX28.

Row 32 contains the cross-spectrum of signals #1 and #2. The formula in Cell A32 is @IMPRODUCT(@IMCONJUGATE(A10),A20). This is copied to Cells A32..DX32.

Row 34 contains the un-normalized cross-spectrum magnitude. The formula in Cell A34 is @IMABS(A32). This is copied to Cells A34..DX34.

Row 38 contains the coherence or MSC function. The formula in Cell A38 is +A34^2/(A14*A26)/1024. The factor 1024 is for normalization. This formula is copied to Cells A38..DX38. See Fig. 10-12.

Row 40 contains the SNR function. The formula in Cell A40 is +A38/(1–A38). This is copied to Cells A40..DX40.

Row 42 contains the SNR function in dB. Cell A42 contains the formula 10*@LOG(A40). This is copied to Cells A42..DX42. See Fig. 10-13.

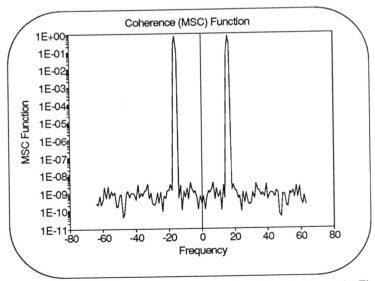

Fig. 10-12. Coherence or MSC function for the signals shown in Fig. 10-11. It is clear that there is considerable narrow-band coherence.

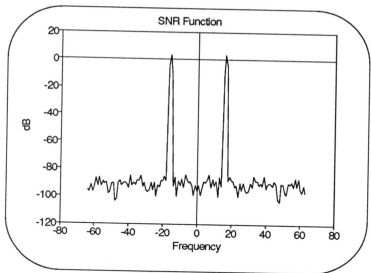

Fig. 10-13. SNR function in dB for the signals shown in Fig. 10-11. The SNR is very small except for a narrow band.

Row 44 contains the centered frequency axis of 128 points, ranging from −64 to +64, including zero.

Row 46 contains the peak detector. The formula contained in Cell A46 is @IF(BM40 =$B$3,BM44,0). If this is the maximum value at a positive frequency then it writes the frequency (BM44) at which the maximum occurs. If it is not the maximum value this writes a zero in A46. This formula is copied to Cells A46..BN46.

System Identification

What Is It?

We have used the system function $H(f)$ and impulse response $h(t)$ before so it should be familiar. We found that they were related by the Fourier transform,

$$H(f) = FFT[h(t)] .$$

Expressed in terms of $S_2(f)$ and $S_1(f)$ we can see that $H(f)S_1(f)$ is a cross-spectrum.

$$S_2(f) = H(f)S_1(f) .$$

In system identification we would like to measure $H(f)$ by means of the output and input spectra,

$$H(f) = \frac{S_2(f)}{S_1(f)} .$$

Another interesting and sometimes useful description of $H(f)$ can be obtained in terms of the cross-power spectrum of the input and output, divided by the power spectrum of the input. To see this, simply multiply the equation above as follows,

$$H(f) = \frac{S_2(f)}{S_1(f)} \frac{S_1^*(f)}{S_1^*(f)} = \frac{G_{2,1}(f)}{G_{1,1}(f)} .$$

What Does It Tell You?

Clearly, system identification reveals the system function and impulse response, the heart and soul of the system. It is well known that a cryptographic system can usually be broken if the code-breaker can put his or her own signal into the system and get it back. This is one aspect of system identification using a known input and a measured output.

System Identification Tips

Be careful! System identification can be an unstable operation. Avoid using input signals that have zeros in their spectra. On the other hand, if you are adventurous you can take comfort in the fact that there is small probability of hitting an absolute zero in the spectrum. Adding a little random noise to the input is often helpful in eliminating the zeros.

Random noise alone makes a good test signal. A safe-cracker twiddles the dial of the safe (a random input) and listens and feels for the clicks (the output). Use your knowledge of system identification for good, not evil.

IDENTIFY, System Identification Worksheet

Worksheet Organization

The Home screen of this worksheet is shown in Fig. 10-14.

Row 4 is the Time axis. On the diskette this goes from A4..DX4 in increments of 1, for a total of 128. As usual the number of data points in the row must be in the 2^n form.

Row 6 is the input signal (waveform #1). On the diskette the test signal in an impulse or delta function located in cell A6. There are zeros in the rest of this row. Experiment with different test signals such as sine waves, pulses, pulse groups, and random noise.

Row 8 is the checkerboarded input signal for centering the spectrum. The formula in Cell A8 is +A6*@COS(@PI*A4). This is copied to Cells A8..DX8.

Row 10 is the input spectrum S_1, the FFT of row 8.

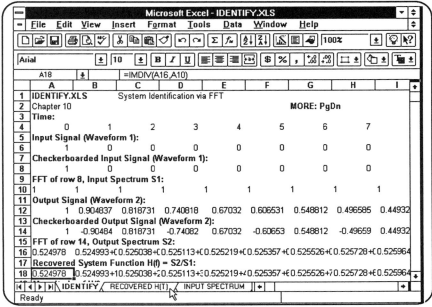

Fig. 10-14. Home screen of IDENTIFY. Cell A18 shows use of @IMDIV. The test signal here is a delta function in Cell A6 which produces the uniform spectrum in row 10.

Row 12 is the output signal, which you would normally measure and place in this row. On the diskette there is an output signal of the form @EXP(–A4/10). This is copied to Cells A12..DX12. See Fig. 10-15.

Row 14 is the checkerboarded output signal. The formula in Cell A14 is +A12*@COS(@PI*A4). This is copied to Cells A14..DX14.

Row 16 is the output spectrum S_2, the FFT of row 14.

Row 18 is the system identification in terms of the system function, $H(f)=S_2/S_1$. The formula in Cell A18 is @IMDIV(A16,A10). This is copied to Cells A18..DX18.

Row 20 is the magnitude of the system function. The formula in Cell A20 is @IMABS(A18). This is copied to Cells A20..DX20. See Fig. 10-16.

Row 22 is the phase of the system function. The formula in Cell A22 is @DEGREES(@IMARGUMENT(A18)). This is copied to Cells A22..DX22.

Row 24 is the IFFT of the system function. This usually has a small imaginary part (an error) even when the impulse response is completely real.

Row 26 removes the imaginary error, recovering the system-identified impulse response. The formula in Cell 26 is @IMREAL(A24). This is copied to Cells A26..DX26. See Fig. 10-17.

Row 32 contains the spectrum of S_1.

Row 34 contains the phase of S_1.

Row 36 is the centered frequency axis, used for graphing.

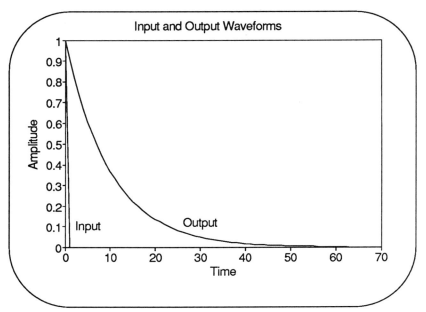

Fig. 10-15. The impulse (delta function) input and the impulse response (output).

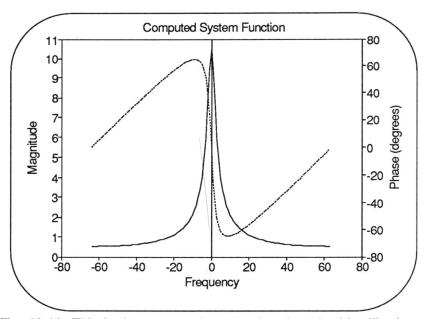

Fig. 10-16. This is the computed system function, the identification of the system. Note the phase at the ends suffers from FFT circularity but the magnitude is small there. Here, system identification is best at low frequencies.

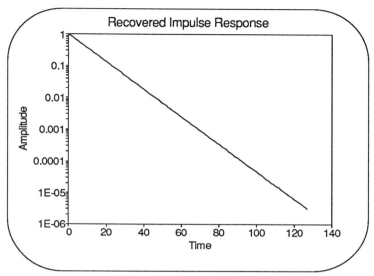

Fig. 10-17. The computed impulse response, the IFFT of Fig. 10-16. Note how accurately the exponential decay is recovered. Try this with other test signals, including a sample of random noise.

Cepstrum

What Is It?

The cepstrum is the inverse Fourier transform of the log of the spectrum magnitude, defined in terms of the IFFT as,

$$C(t) \;=\; \mathrm{IFFT}\; \log[|S(f)|]$$

What Does It Tell You?

The convolution theorem informs us that when signals are convolved in the time domain their spectra are simply multiplied. Taking the log of this spectral product converts it into a sum, and taking the IFFT transforms back to the time domain. This is often helpful in separating echos or "multi-path" due to reflections. After deleting the apparent echos (with a "short-pass" filter) in the time domain (after IFFT) you can use the FFT to get back to the frequency domain. Next, take the anti-log of the spectra, and then use the IFFT to get back to the time domain with the echos removed.

This process is sometimes called homomorphic deconvolution because the desired signal and its echos are similar in form. Note the curious analogy in time domain operations and frequency domain operations. The short-pass filter passes signals which occur early and removes signals with long time delays. This is analogous to the familiar low-pass filter in the frequency domain.

The process can also be carried out for the converse case in which two spectra are convolved. This means that their corresponding time functions are simply multiplied. Performing the cepstrum operation will separate the convolved spectra after the log operation in the time domain. Clearly this can be performed with more than two signals.

CEPSTRUM, Cepstrum Worksheet

In CEPSTRUM we will generate a signal and its echo that are convolved in the time domain, and then we will recover the original signal using the cepstrum and short-pass filtering. The echo could be recovered using long-pass filtering.

Worksheet Organization
The Home screen is shown in Fig. 10-18. *Rows 4 - 18 produce the convolved, noisy test signal.*

Row 4 is the Time axis. On the diskette this goes from A4..DX4 in increments of 1, for a total of 128. As usual the number of data points in the row must be in the 2^n form.

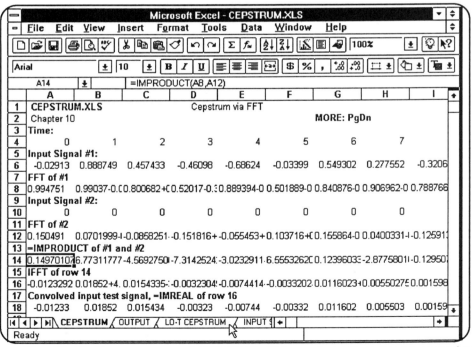

Fig. 10-18. Home screen of CEPSTRUM. This worksheet removes convolved echos from the original signal. The first 18 rows produce the test signal.

Row 6 is input waveform #1. On the diskette the installed test signal is @EXP(-A4/11)*@SIN(2*@PI*0.2*A4)+(@RAND*0.1–0.05). This is copied to Cells A6..DX6.

Row 8 is the FFT of row 6. No checkerboarding is used because we're not interested in the centered spectrum.

Row 10 is input waveform #2. This signal starts in Cell W10; there are zeros from A10..V10. In the diskette file the installed test signal in Cell W10 is @EXP(-A4/11)*0.22*@SIN(2*@PI*0.2*A4). This is copied to Cells W10..DX10.

Row 12 is the FFT of row 6. Again, no checkerboarding is used.

Row 14 is the cross-spectrum. Cell A14 has @IMPRODUCT(A12, A8). This is copied to Cells A14..DX14.

Row 16 is the IFFT of row 14.

Row 18 is the input test signal, the convolved signals #1 and #2. The formula in Cell A18 is @IMREAL(A16). This is copied to Cells A18..DX18. See Fig. 10-19.

Row 22 is the FFT of the input test signal from row 18.

Row 24 uses @LN to separate the spectral product. The formula in Cell A24 is @LN(@IMABS(A22)). This is copied to Cells A24..DX24.

Row 26 is the IFFT of row 24.

Row 28 is the real part of the cepstrum. The formula in Cell A28 is @IMREAL(A26). This is copied to Cells A28..DX28. See Fig. 10-20

Row 30 is the low-time filtered version of row 28. The formula in Cell A30 is +A28. This is copied to Cells A30..V30. Put 0 in cell W30 and copy this to Cells W30..DX30. See Fig. 10-21.

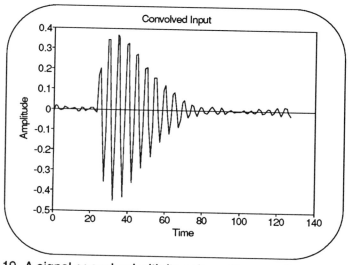

Fig. 10-19. A signal convolved with its echo, with some time overlap. This is a multi-path problem in communications, radar, sonar, architectual acoustics, and fiber optics. Often multiple echos will appear as system noise.

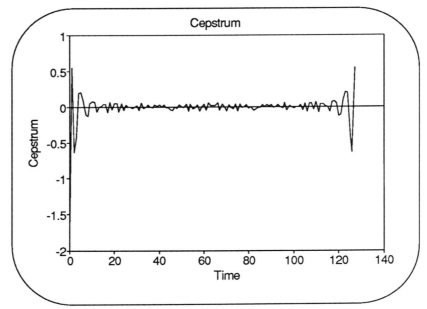

Fig. 10-20. The cepstrum before time filtering. This is not readily interpreted.

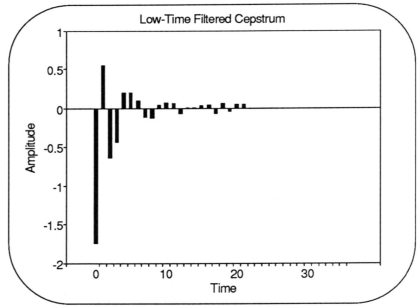

Fig. 10-21. The cepstrum after short-pass filtering. This must be done by trial and error, observing the final result, because it is not clear where an echo occurs.

Row 32 is the FFT of row 30.

Row 34 restores the spectral product by exponentiating row 32. The formula in cell A34 is @IMEXP(A32). This is copied to Cells A34..DX34.

Row 36 is the IFFT of row 34. This takes you back to the original time domain with the low-time filtered cepstrum. Some people call time-filtering "liftering."

Row 38 removes any residual imaginary part. The formula in Cell A38 is @IMREAL(A36). This is copied to Cells A38..DX38. See Fig. 10-22.

It seems that there is no better method yet known for short-pass and long-pass filtering than to cut and try. You can also conceive of time-pass and time-reject filtering in analogy with band-pass and band-reject filters in the frequency domain. If the echos in the cepstrum are easily identified this simplifies the matter but it still must be done by eye and hand. Using the complex cepstrum permits recovery of phase information.

Analytic Signals

What Are They and Who Needs Them?

To this point we've used signals that can be represented by completely real functions. However, we found that real signals have complex Fourier transforms. What if the signals are complex? The use of complex signals can sometimes be helpful because of the simplification and the new dimension that arises. At first this may sound paradoxical. Read on. (We will use complex signals later in phased arrays.)

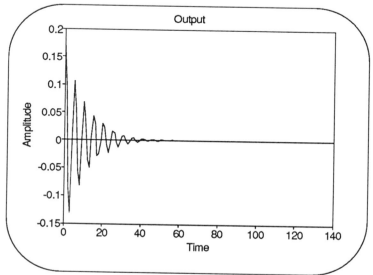

Fig. 10-22. The signal after removal of the convolved echo. Compare Fig. 10-19.

Consider a completely real signal, an oscillatory exponential decay such as we've used before,

$$s(t) = A \, e^{-kt} \cos(2\pi ft) \ .$$

As we have seen many times, the spectrum of such a function has both positive and negative frequencies and the spectrum is symmetric. If you know the spectrum for positive frequencies you also know the spectrum for negative frequencies. The Fourier transform is redundant. If you want to know the total energy just double the energy in the positive range. Now let's consider a complex signal represented by

$$s(t) = A \, e^{-kt} \, [\cos(2\pi ft) + j\sin(2\pi ft)]$$

$$s(t) = A \, e^{-kt} e^{j2\pi ft}$$

$$s(t) = A \, e^{-kt + j2\pi ft}$$

$$s(t) = A \, e^{-(k - j2\pi f)t}$$

This is called an *analytic signal*. It is easy to verify by the FFT or by means of manual calculation that this complex signal has only positive frequencies. The whole thing looks like an exponential decay with the factor $(k - j2\pi f)$ as a complex decay constant.

Does this seem unnecessarily complicated? It is true that you can get along without complex functions in many walks of life. Even our beloved Fourier transform can be expressed completely in terms of real functions; this is called the Hartley transform. Probably the only time you are compelled to use complex functions is in quantum mechanics because real functions are not solutions of Schrödinger's equation. But that's another story.

Not all complex signals are analytic. There are many tests for analyticity but I promised you we wouldn't use any calculus. To make a real signal into an analytic signal you can use the Hilbert transform because the real and imaginary parts of an analytic signal are related by this transform. If you know the real part you can calculate the imaginary part, and conversely. Causality is an integral part of the Hilbert transform. There is an extensive literature concerning analytic functions and the part they play in mathematics, engineering, and physics.

Hilbert Transform via FFT

If you're at a party and things get dull you can use this to amuse and entertain your friends. (If they don't invite you again, don't blame me.) This is an easy way to convert a real signal into an analytic signal:

(a) Use the FFT to obtain the complex spectrum of the real signal.

(b) Exchange the real and imaginary parts.

(c) Change the sign of the new real part for f less than 0.

(d) Change the sign of the new imaginary part for f greater than 0.

(e) Use the IFFT on the new spectrum to obtain the imaginary part of the analytic signal.

(f) Add the real and imaginary parts to form the analytic signal.

Clearly, you can use this to test a complex signal to see if it is analytic. Run the real part through operations (a) - (f) and if you get back the original complex signal then you know that it's analytic. Some real signals cannot be made analytic because they do not have Hilbert transforms.

ANALYTIC, Analytic Signal Worksheet

The Home screen is shown in Fig. 10-23. Input new signals in rows 6 and 8.

Row 4 is the Time axis. We have chosen 128 points on the diskette worksheet for convenience in doing the FFT later.

Row 6 is the real part of the input signal. The formula contained in Cell A6 is @EXP(–A4*0.05)*@COS(A4*0.5). This copied to Cells A6..DX6.

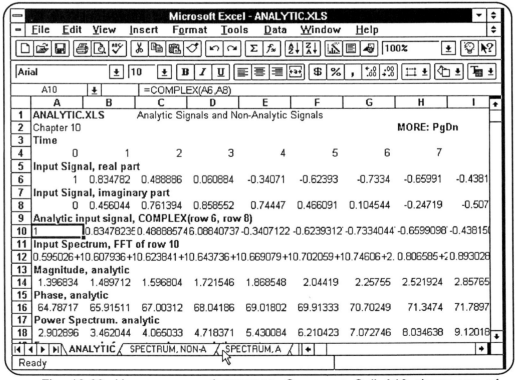

Fig. 10-23. Home screen of ANALYTIC. Cursor on Cell A10 shows use of @COMPLEX to form analytic signal from real and imaginary parts.

Row 8 is the imaginary part of the input signal. The formula in Cell A8 is @EXP(-A4*0.05)*@SIN(A4*0.5). This is copied to Cells A8..DX8.

Row 12 is the FFT of row 10.

Row 10 is the analytic signal. The formula in cell A10 is @COMPLEX(A6,A8). This is copied to cells A10..DX10. The polar plot is shown in Fig. 10-27.

Row 14 is the magnitude of the spectrum in row 10. The formula in Cell A14 is @IMABS(A10). This is copied to Cells A14..DX14. See Fig. 10-25.

Row 16 is the phase of the spectrum in row 10. The formula in Cell A16 is @DEGREES(@IMARGUMENT(A10)). See Fig. 10-25.

Row 18 is the power spectrum. The formula in Cell A18 is +A14^2. This is copied to Cells A18..DX18. Standard normalization would be to divide each Cell by 128.

Row 20 is the centered frequency axis for the non-analytic signal. There are 128 points from −64 to +63 in steps of 1.

Row 24 is the FFT of the real (non-analytic) signal, row 6.

Row 26 is the magnitude of row 24. The formula in Cell A26 is @IMABS(A24). This is copied to Cells A26..DX26. See Fig. 10-26.

Row 28 is the phase of the non-analytic signal. The formula in Cell A28 is @DEGREES(@IMARGUMENT(A24)). This is copied to Cells A28..DX28.

The time evolution of the example analytic signal is shown in Fig. 10-24. This function has only positive frequencies and can be represented as a rotating vector in Fig. 10-27.

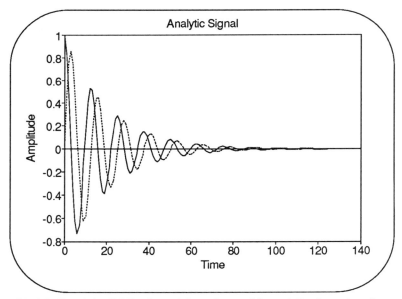

Fig. 10-24. Real (solid line) and imaginary (dotted line) parts of an analytic signal. Spectral magnitude (solid line) and phase (dotted line) are shown below.

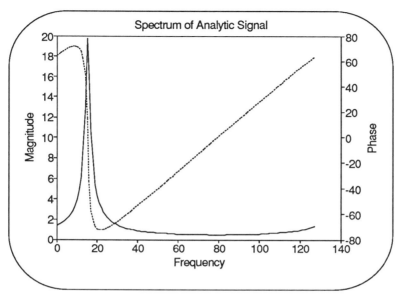

Fig. 10-25. Spectrum of the analytic signal shown in Fig. 10-24. Because of FFT circularity only the first 64 points on the frequency axis should be considered. Ordinarily the graph would be plotted from zero to 63. Rectangular window.

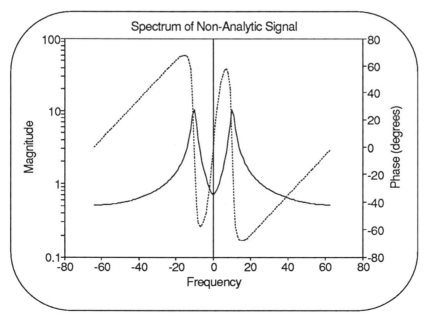

Fig. 10-26. The spectrum of the non-analytic signal contains both positive and negative frequencies, as shown in this FFT with a rectangular window.

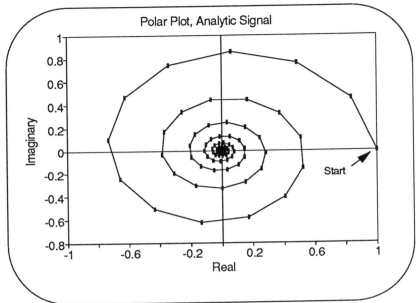

Fig. 10-27. The rotating vector nature of the analytic signal is evident in the polar plot, looking back along the time axis toward the origin. Time increment between markers is constant.

Phased Arrays

What Are They?

Phased arrays are ensembles of coherent sources or transducers such as microphones, loudspeakers, hydrophones, and antennas. The radiation pattern of each source or transducer combines in amplitude and phase by the principle of superposition. The resultant radiation pattern can be much sharper than an individual source and, just as important, the beam can be steered by electronic phase shifters at a very rapid rate to point in different directions. In addition, the beam pattern can be designed to have sharp nulls so an interfering, undesired signal can be minimized during reception. Conversely, during transmission the sharp nulls will minimize radiation in those directions. These techniques find wide use in radar and sonar.

The phased array is an ideal antenna for computer control of the outgoing signal and the incoming signal. Electromagnetic phased arrays have an additional dimension of polarization which acoustic arrays lack. Polarization provides even more flexibility in sophisticated radars.

Human hearing is based on an array of two transducers which permits elementary direction finding when they are connected to a good computer like the brain. At low frequencies direction finding is done by phase difference and at higher frequencies intensity difference is used. Human direction finding by phase difference is not very good at wavelengths less than twice the ear-to-ear distance. This is true for any phased array. The beam pattern rapidly deteriorates when the element spacing exceeds a half wavelength. Also, it might be better if people had 7 or more ears. You will be able to experiment with the number of ears and their spacing in the PHASE-RA worksheet.

Everything we have done before with the FFT is applicable to a one-dimensional phased array. Now we can go a little farther and choose the amplitude *and phase* of each point in the signal, which is essentially what we did in the ANALYTIC worksheet for each point as a function of time. It is not difficult to extend this method to two-dimensional phased arrays which let you swing the beam in two dimensions.

This worksheet has two independent parts. In the first part, shown in the Home screen in Fig. 10-28, the beam pattern is computed by superposition. In this method you can choose the element spacing, wavelength, number of elements, and beam deflection (left or right) measured from the forward direction. For simplicity all elements have the same amplitude in this section. This part of PHASE-RA does not require the FFT.

The second section of the worksheet (Fig. 10-33) uses the FFT and lets you choose all of the above parameters plus you can experiment with different windows to adjust the power (or sensitivity) of each element to control sidelobes as we did before in time and frequency. These things will become clear as you use the worksheet.

EXAMPLE:

An adult human can locate a low-frequency sound (500 to 800 Hz) to within about 10 degrees. Experiments show that human direction-finding is influenced by both time delay (phase shift between ears) and intensity difference, with maximum error occurring between 3 and 4 kHz. At high frequencies intensity appears to be the dominant factor. Suppose an adult has an ear-to-ear distance of 20 cm and the speed of sound is 345 m/s. A half wave at this speed corresponds to a frequency of,

$$f = \frac{v}{\lambda} = \frac{345 \ \text{m/s}}{2 \times 0.2 \ \text{m}} = 862.5 \ \text{Hz} \ .$$

Above this frequency the two-element array of the ear-brain combination is not very good at directions using phase shift, which is why intensity information is employed at high frequencies. Acoustic "shadowing" by the head complicates the analysis. Try this simple experiment in the following worksheet for two elements: Start with a spacing of 0.5 λ and increase the spacing in steps of 0.1 λ. The change in the polar plot is dramatic.

Acoustic direction-finders have been used for military purposes for hundreds of years. Before electronics these instruments were constructed in the form of two large horns, often separated by 2 m or more. This separation is necessary to localize cannon noise which has a lot of energy at low frequencies, far away from the source.

PHASE-RA, The Phased Array Worksheet

This worksheet computes the radiation pattern of a one-dimensional phased array, using two methods. First, the intensity distribution is computed by simple superposition of multiple sources. Second, the FFT is used to compute the radiation pattern. You will see that the FFT provides considerable flexibility in adjustment of parameters, such as use of our familiar data windows to modify the beam distribution and control sidelobes.

For the Home screen (Fig. 10-28) the beam pattern is computed for N identical elements with a spacing of d and a wavelength of λ. The beam deflection θ_o is measured from the forward direction and the phase shift between adjacent elements is φ. The equation for the beam distribution [or Gain $G(\theta)$] shows the importance of the dimensionless parameter d/λ,

$$G(\theta) = \frac{\sin^2 [N\pi(d/\lambda)(\sin\theta - \sin\varphi)]}{N^2\sin^2 [\pi(d/\lambda)(\sin\theta - \sin\varphi)]} \ .$$

In this equation the phase shift φ between elements is given by,

$$\varphi = 2\pi(d/\lambda) \sin\theta_o \ .$$

The maximum of the beam occurs when $\theta = \theta_o$. As you experiment with this part of the worksheet you will discover that the beam pattern becomes sharper as the number of elements increases, but the number of sidelobes also increases. This is compensated to some extent by the fact that the relative intensities of the sidelobes decrease. You will also discover that the beam width θ_B (measured by the half-power points in angle) increases as the deflection increases. The valid range (sine ≤ 1) of the equation below increases as N increases and it is always more accurate at small beam deflections,

$$\theta_B = \sin^{-1} [0.443(\lambda/Nd) + \sin\theta_o] + \sin^{-1} [0.443(\lambda/Nd) - \sin\theta_o].$$

Worksheet Organization

Cell A4 is d, the element spacing. Enter this value in meters.

Cell A6 is λ, the wavelength. Enter this value in meters.

Cell A8 is N, the number of elements. Enter this value (choose 1 through 128).

Cell A10 is the desired angle of beam deflection, in degrees. Enter this value. (Choose any value from –90 to +90. The beam pattern is best between about ±60 degrees.)

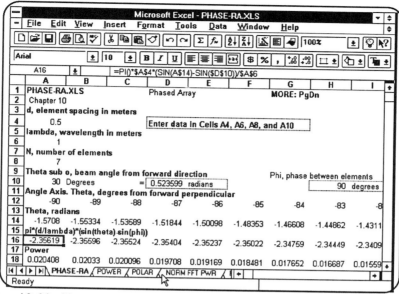

Fig. 10-28. Home screen of PHASE-RA. This screen shows the part of the worksheet that computes the beam pattern by superposition of elements.

Cell D10 converts the beam deflection to radians. The formula in cell D10 is +A10*@PI/180.

Cell H10 computes the relative phase shift between elements. The formula in cell H10 is 2*@PI*(A4/A6)*@SIN(D10)*180/@PI.

Row 12 is the angle axis in degrees. Use the Fill command to set the range and increment. At present this row goes from –90 to +90 with an increment of one. You can change this to zoom in on a small range of angles.

Row 14 is the angle axis in radians. The formula in Cell A14 is +A12*@PI/180. This is copied to Cells A14..DX14.

Row 16 is the angle φ. Cell A16 contains the formula,
@PI*($A$4* (@SIN($A$8*(A16 +1E-08))^2/(@SIN(A16+1E-08)^2).

The factor 1E-08 is included to eliminate round-off error in @SIN at zero.

Row 18 computes the power distribution of the beam. The formula in Cell A18 is [(1/$A$8^2)*@SIN($A$8*(A16 +1E-08))^2]/@SIN(A16+1E-08)^2. This is copied to Cells A18..DX18.

Row 20 scales the power distribution in dB. The formula in Cell A20 is 10*@LOG(A18). This is copied to Cells A20..DX20. See Figs. 10-29 and 10-31. Row 22 is used for a polar plot. See Figs. 10-30 and 10-32.

This computes the x-projection of power. The formula contained in Cell B22 is +B18*@COS(B14). This is copied to Cells B22..DX22.

Row 24 is used for a polar plot. This computes the y-projection of power. The formula in Cell A24 is +A18*@SIN(A14). This is copied to Cells A24..DX24.

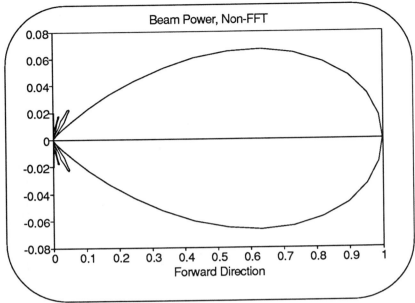

Fig. 10-29. Polar plot of a 7-element array, equal power, all in phase, d/λ = 0.5. Note sidelobes. Scale on *y*-axis is expanded to show details of beam pattern.

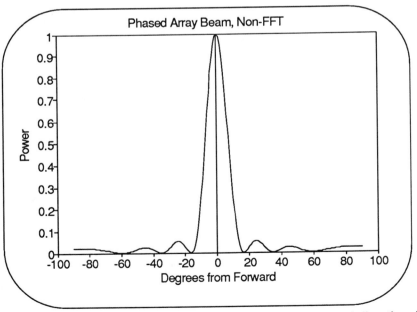

Fig. 10-30. Beam power as a function of angle from forward direction. This is the same information that is plotted in Fig. 10-29 in polar format.

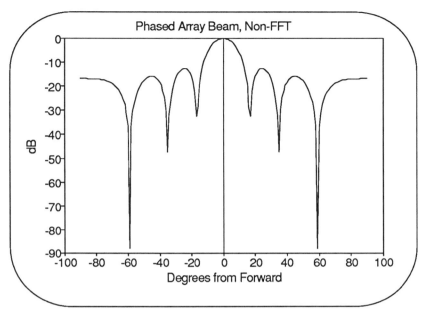

Fig. 10-31. This is the same information shown in Figs. 10-29 and 10-30, but here it is scaled in dB. Each null would ideally approach minus infinity.

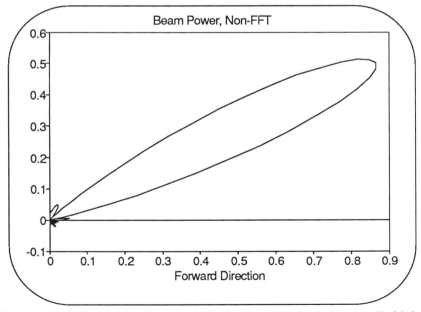

Fig. 10-32. Polar plot for a beam deflection of +30°. Note the small sidelobes at the origin. This result is for seven elements with equal power, $d = \lambda/2$.

FFT Method

This section of PHASE-RA provides an alternative method of computing the beam pattern. In rows 31 and 39 the value of d/λ has been set to 0.5. See Fig. 10-33. for the Home screen.

Cell D29: Enter the beam deflection in degrees.

Cell F29: This calculates the beam deflection in radians. The formula in cell F29 is @RADIANS(D29).

Row 31 is the array axis in units of d/λ. On the diskette this axis has a range of 0..63.5, for a total of 128 points. You can adjust the increment and range to zoom in on the pattern, but remember to use 2^n points which is required by the FFT.

Row 33 contains the element amplitudes. You can adjust these as you wish. On the diskette there are 1's in Cells A33..G33 and 0's from H33..DX33.

Row 35 contains the weight or window function. On the diskette the window is all 1's (uniform or rectangular window). You should experiment with some different windows and observe their effects in reducing sidelobes. This is analogous to our previous use of data windows in enhancing frequency resolution.

Row 37 produces the checkerboarded and weighted element amplitudes. The formula in Cell A37 is +A35*@COS(2*@PI*A31)*A33. This is copied to Cells A37..DX37.

Row 39 sets the phase in degrees between each element, starting with the element at 0 on the array axis. Cell A39 contains the value 0. The formula in Cell B39 is +A39 +@DEGREES(2*@PI*0.5*@SIN($F$29)). This is copied to Cells B39..DX39.

Row 41 computes the real part of each element. The formula contained in A41 is +A37*@COS(@RADIANS(A39)). This is copied to Cells A41..DX41.

Row 43 computes the imaginary part of each element. The formula in A43 is +A37*@SIN(@RADIANS(A39)). This is copied to Cells A41..DX41.

Row 45 produces the complex, weighted, and checkerboarded elements. The formula in A45 is @COMPLEX(A41,A43). This is copied to Cells A45..DX45.

Row 47 is the FFT of row 45. Remember to reccalculate the FFT whenever you change an array parameter because it does not recalculate automatically.

Row 49 computes the magnitude of the beam distribution. The formula in Cell A49 is @IMABS(A47). This is copied to Cells A49..DX49.

Row 51 computes the phase of the beam distribution. The formula in Cell A51 is @DEGREES(@IMARGUMENT(A47)). This is copied to Cells A51..DX51.

Row 53 computes the power in the beam distribution. The formula in Cell A53 is +A49^2. This is copied to Cells A53..DX53.

Row 55 scales the power in dB. The formula in A55 is 10*@LOG(A62). This is copied to Cells A55..DX55. (Cell $B$60 is the maximum power.)

Row 57 is used for the polar plot, x-component. Cell A57 contains the value 0 to avoid roundoff error. The formula contained in Cell B57 is +B62*@COS(@RADIANS(B66)). This is copied to Cells B57..DX57.

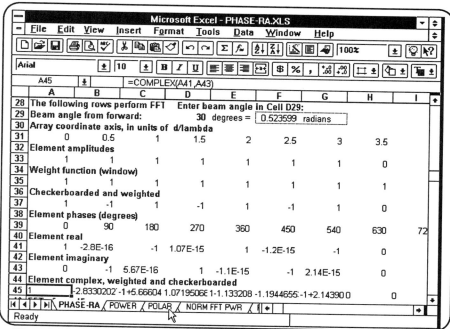

Fig. 10-33. Second screen of PHASE-RA, showing first part of the FFT section. Cell A45 shows use of @COMPLEX in setting element amplitude and phase.

Row 58 is also used for the polar plot, y-component. Cell A58 contains the value 0. The formula in Cell B58 is +B62*@SIN(@RADIANS(B66)). This is copied to Cells B58..DX58.

Cell B60 detects the maximum power, used in dB scaling. The formula in this Cell is @MAX(A53..DX53).

Row 62 normalizes the power. The formula in Cell A62 is +A53/$B$60. This is copied to Cells A62..DX62.

Row 64 contains the centered angle axis before conversion to linearized angle measure. On the diskette this axis has a range of −32 to 31.5 in steps of 0.5. You can adjust this to zoom in on a shorter range but always use 2^n points.

Row 66 is the centered angle axis, linearized in degrees. The formula in Cell A66 is @DEGREES(@ASIN(A64/32)). This is copied to Cells A66..DX66. Note the use of the inverse sine function for proper axis scaling.

Experiment with the number of elements, spacing, and wavelength. It is evident that phased arrays are useful for producing a narrow, rapidly steerable beam. With a two-dimensional array the beam can be steered in two dimensions. In addition to active and passive radar and sonar a phased array can be used as a high-power microwave system composed of numerous low-power devices to produce a very intense, highly localized beam.

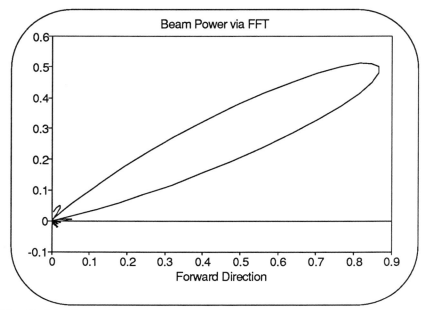

Fig. 10-34. Phased array beam for a deflection of +30°, seven elements, equal power, half-wave spacing. Compare with Fig. 10-32 for non-FFT computation.

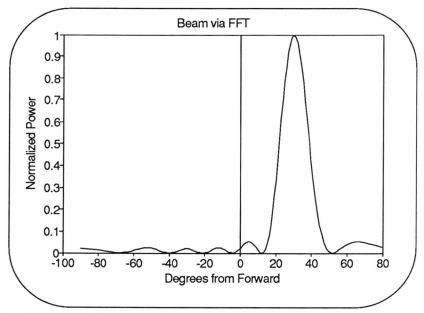

Fig. 10-35. This is the same information shown in Fig. 10-34, but a different presentation.

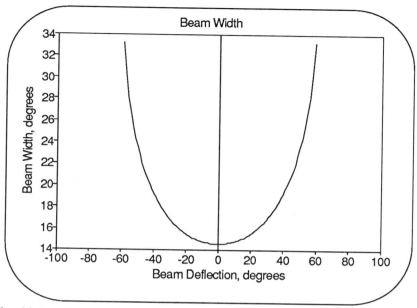

Fig. 10-36. Beam width (half-power points in angle) for 7 elements with half-wave spacing and equal power.

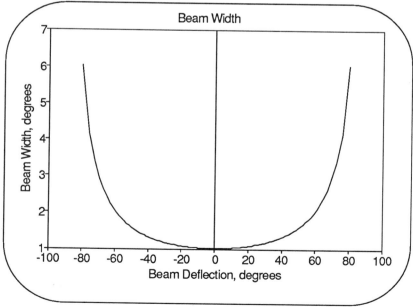

Fig. 10-37. Beam width for 100 elements with half-wave spacing and equal power. The AN/FPS-108 Cobra Dane radar has 15,360 active elements.

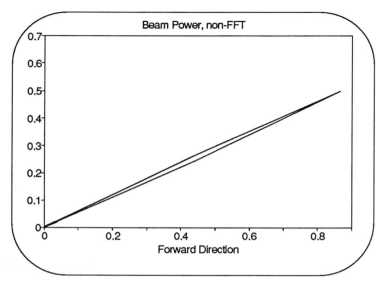

Fig. 10-38. Beam power for 58 elements, 30° deflection, half-wave spacing, and equal power. The polar plot for 100 elements (corresponding to Fig. 10-37) is almost a straight line. Compare with Figs. 10-32 and 10-34 for 7 elements.

Experiment with the worksheet to discover how small errors in power and phase of each element affect the beam. Also investigate the effect of one or more element's total failure on the beam pattern; element failure is less serious for large N.

Digging Deeper

Engineering Applications of Correlation and Spectral Analysis, 2nd Ed., J. S. Bendat and A. G. Piersol (John Wiley & Sons, New York, 1993). ISBN 0-471-57055-9. There are good discussions of correlation and spectral density functions, system identification, and propagation-path identification, among many others.

Digital Signal Processing, A. V. Oppenheim and R. W. Schafer (Prentice-Hall, Englewood Cliffs, New Jersey, 1975). ISBN 0-13-214635-5. This is a comprehensive introduction to the mathematical foundations of digital signal processing, at the senior and graduate level. Chapter 7 covers the discrete Hilbert transform; Chapter 10 has a good treatment of the complex cepstrum and homomorphic deconvolution.

Theory and Application of Digital Signal Processing, L. R. Rabiner and B. Gold, (Prentice-Hall, Englewood Cliffs, New Jersey, 1975). ISBN 0-13-914101-4. Chapter 2 has the discrete Hilbert transform and Chapter 12 discusses speech processing.

SSP: The Spreadsheet Signal Processor, S. C. Bloch (Prentice-Hall, Englewood Cliffs, New Jersey, 1992). ISBN 0-13-830506-4. Chapters 4 and 13 discuss cross-correlation in the original data domain. Chapter 11 treats system identification in the original data domain. Two diskettes are included.

Advanced Topics in Signal Processing, J. S. Lim and A. V. Oppenheim, Eds. (Prentice-Hall, Englewood Cliffs, New Jersey, 1988). ISBN 0-13-013129-6. In the *potpourri* of interesting material here, Chapter 6 (by S. H. Nawab and T. F. Quatieri) discusses the short-time Fourier transform in speech, music, and sonar processing. This type of joint-time-frequency analysis is also useful in Doppler processing when the frequency is shifting.

Handbook of Digital Signal Processing, D. F. Elliott, Ed., (Academic Press, San Diego, California, 1987). ISBN 0-12-237075-9. Chapter 11, by M. T. Silva, has a good introduction to time delay estimation and phased arrays.

Introduction to Random Signals and Applied Kalman Filtering, R. G. Brown and P. Y. C. Hwang (John Wiley & Sons, New York, 1992). ISBN 0-471-52573-1. A diskette is included. There is extensive discussion of applications to the Global Positioning System.

Fundamentals of Complex Analysis for Mathematics, Science, and Engineering, Second Ed., E. B. Saff and A. D. Snider (Prentice-Hall, Englewood Cliffs, New Jersey, 1993) ISBN 0-13-327461-6. Chapter 2 treats analytic functions, and Chapter 8 has the Fourier and Hilbert transforms.

Signal Processing Algorithms, S. D. Stearns and R. A. David (Prentice-Hall, Englewood Cliffs, New Jersey, 1988). ISBN 0-13-809435-7. A diskette is included; all algorithms are expressed in FORTRAN 77.

Introduction to Radar Systems, M. I. Skolnik (McGraw-Hill, New York, 1962). Library of Congress Catalog Card Number 61-17675. Chapter 7 has a discussion of phased arrays and antenna pattern synthesis.

The Fast Fourier Transform and Its Applications, E. O. Brigham (Prentice-Hall, Englewood Cliffs, New Jersey, 1988) ISBN 0-13-307505-2. Chapter 14 introduces the use of the FFT in antenna analysis and synthesis.

The Fourier Transform and Its Applications, Second Ed., R. N. Bracewell (McGraw-Hill, New York, 1986). ISBN 0-07-007015-6. Chapter 13 treats antenna patterns with the Fourier transform.

Chapter 11

Signals, Noise and Decibels

What This Chapter is About

In previous chapters we have mostly considered ideal, noiseless signals. In the real world, signals are accompanied by a random component that we call noise. In addition to random noise there may also be a nonrandom component consisting of undesired signals (interference). In this chapter we will only discuss the noise.

Thermal (Nyquist, Johnson," White") Noise

The random motion of electric charges constitutes a random current, and this random current produces random voltages. Because the random motion, like Brownian motion, increases with temperature, the random voltage increases with temperature. This was investigated theoretically by H. Nyquist and experimentally by J. B. Johnson, so this type of noise is sometimes called *Nyquist noise* or *Johnson noise*. It is also known as *white noise*, because its energy is the same at all frequencies .

You can't get away from thermal noise because every conductor and semiconductor has charge carriers, but thermal noise can be reduced by decreasing temperature, decreasing bandwidth, and decreasing resistance. The open-circuit rms thermal voltage across the terminals of a resistor is given by,

$$V = \sqrt{4kTBR} \ .$$

In this equation V is the rms noise voltage in volts, k is Boltzmann's constant $(1.38 \times 10^{-23}$ Joule/K), T is the absolute temperature on the Kelvin scale, B is the bandwidth in Hz, and R is the resistance in Ohms.

It is obvious that thermal noise can be reduced by decreasing bandwidth and the resistance. At first it may appear attractive to use a narrow filter to reduce the bandwidth but this limits the rate of information transmission. In many cases limiting bandwidth is not a viable option. Many applications, like FM, high data rate, and spread-spectrum systems, require large bandwidths for their operation.

Examples:

- What is the open-circuit voltage (rms) across a 1kΩ resistor at 300K (room temperature)? The voltmeter has a bandwidth of 100 kHz.

$$V = \sqrt{4(1.38\times10^{-23})(300K)(10^5\ \text{Hz})(10^3\ \Omega)} = 1.3\times10^{-6}\ V = 1.3\ \mu V$$

This is a small voltage but it can be troublesome at the input stage of an audio amplifier. If the amplifier has a voltage gain of 10^6 then the thermal noise will produce a 1.3V signal at the output!

- Now suppose you are using an oscilloscope with a 30MHz bandwidth and the same resistor and temperature as above. What is the rms voltage on the scope?

$$V = \sqrt{4(1.38\times10^{-23})(300K)(3\times10^7\ \text{Hz})(10^3\Omega)} = 2.23\times10^{-7}\ V = 22.3\ \mu V$$

This is a significant voltage, much larger than the sensitivity of input stages of amplifiers in telecommunications systems. These calculations emphasize the point that it is important to keep the input impedance low when you cannot control the temperature in wideband systems.

THERMAL, Thermal Noise Worksheet

The first section of this worksheet lets you input values of bandwidth and resistance, and it calculates the open-circuit rms voltage across the resistor as a function of absolute temperature. The Home screen is shown in Fig. 11-1. The second section, for thermal radiation, can be accessed by pressing PgDn. The second section evaluates the Planck distribution for two user-defined temperatures. The temperature range is also user-defined.

Worksheet Organization

Enter the resistance in Ohms and bandwidth in Hz in Cells E5 and E6, respectively.

Row 9 is the temperature axis. You can adjust the increment and starting point by modifying Cells A9 and B9. In Fig. 11-1 the starting temperature is chosen as 0.1 so you can use the logarithmic scaling available in later versions of most spreadsheets.

Row 12 computes the rms voltage appearing across R at each temperature.

It is useful to compare Fig. 11-2 with Fig. 11-3, in which the same data are plotted with linear scales on both axes. Note that the upper temperature in these Figures is somewhat less than that of the sun's surface.

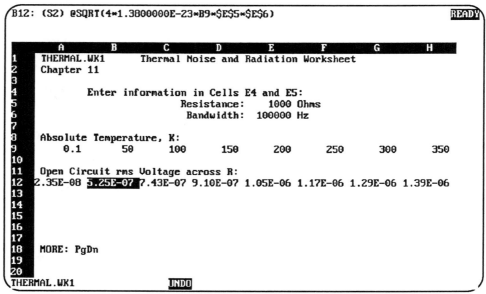

Fig. 11-1. Worksheet for thermal noise. The formula for Cell B12 is displayed.

The graph named *Thermal Noise* in this worksheet is shown in Fig. 11-2, in a log-log format.

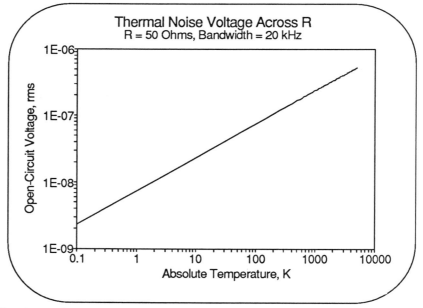

Fig 11-2. Open-circuit thermal noise voltage appearing across a 50 Ohm resistor, for a bandwidth of 20 kHz.

Shot (Schottky) Noise

Whereas thermal noise is due to random motion, shot noise has its origins in the discrete nature of the charge carriers. Shot noise is also called *Schottky noise*, in honor of W. Schottky, who first studied this theoretically. J. B. Johnson did the first careful measurements on this phenomenon. Shot noise is *independent of frequency*, but it depends directly on the bandwidth,

$$I_n^2 = 2qI_{dc}B$$

In the equation for shot noise, I_n^2 is the mean square noise current, q is the electron charge $(1.6 \times 10^{-19}$ coulomb), I_{dc} is the direct current in amperes, and B is the bandwidth in Hz.

Flicker (1/f) Noise

The physical origin of this type of noise is not completely understood, but experiments show that it depends on the flow of current. The name "1/f" noise is due to the fact that this phenomenon increases as the frequency decreases. The mean square voltage associated with 1/f noise can be described empirically as,

$$<V^2> = K\frac{I^2}{f}$$

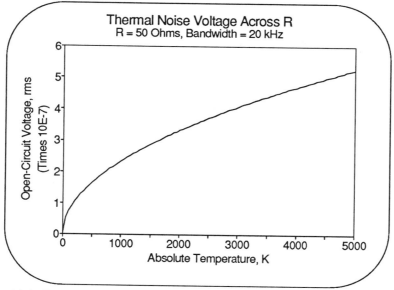

Fig. 11-3. This is the same data shown in Fig. 11-2, but on linear scales.

In this equation K is an empirical constant that depends on the composition and geometry of the resistance and possibly other factors. The dc current in the resistance is I and f is the frequency.

This type of noise is important in semiconductor devices where it may be the *dominant effect* at low frequencies. Noise in transistors consists of thermal, shot, and flicker.

Noise Factor and Noise Figure

The *noise factor* (F) is a common parameter of merit for electronic devices, and is defined as,

$$F = \frac{[\text{Signal Power} / \text{Noise Power}]_{\text{input}}}{[\text{Signal Power} / \text{Noise Power}]_{\text{output}}} .$$

The *noise figure* (NF) is simply the noise factor expressed in dB,

$$NF = 10 \log_{10}(F).$$

When devices are connected in series (cascade) the total noise factor is given by,

$$F_{total} = F_1 + \frac{F_2 - 1}{G_1} + \frac{F_3 - 1}{G_1 G_2} + \frac{F_4 - 1}{G_1 G_2 G_3} + \cdots$$

where F_1 is the noise factor of stage 1 and G_1 is the gain of stage 1 (*not* in dB), and so on. This equation shows a very important fact: If the gain of the first stage is high then *the noise factor and noise figure are primarily determined by the first stage.*

Equivalent Noise Temperature
This is the equivalent temperature T due to the summation of all noise components. The equivalent noise temperature is related to noise figure (in dB) by,

$$NF = 10 \log (1 + T/290)$$

In satellite telecommunications systems, for example, the equivalent noise temperature can be considered to be composed of four parts,

- Passive component noise, the noise produced by passive components before the signal encounters the first active component

- Excess noise from active amplifying stages in the receiver chain

- Leakage noise from high power amplifiers used in the uplink

• Antenna noise, measured at an elevation angle of 5° (the worst case). This noise includes spillover noise from antenna sidelobes, noise from outer space, and atmospheric noise.

Noise from outer space has provided some convincing experimental evidence that our universe started with a bang. While trying to improve satellite communications at Bell Labs, Penzias and Wilson discovered that there seemed to be some nonremovable noise that did not depend on where their antenna was pointed. Cooling their detector did not help. They literally cleaned out their system to remove as much noise as they could. Some of this cleaning included eliminating pigeon droppings from their microwave antenna horn.

During a casual conversation they learned that a group at Princeton, under the direction of P. J. E. Peebles, predicted that, if there had been a Big Bang at the beginning of time, the universe would be filled with radiation left from that event. Furthermore, because of the expansion of the universe and the Doppler shift this fossil radiation would now appear to come from a blackbody radiator at a temperature of about 3 K. An experiment was already being planned by Robert H. Dicke, also at Princeton. Dicke was well qualified for this because he had developed the first microwave radiometer using a lock-in amplifier, a type of analog cross-correlator.

Penzias and Wilson announced that their measurements were consistent with a temperature of 2.7 K at a wavelength of 7.35 cm (4.08 GHz). Additional measurements at many frequencies have shown amazing agreement with theoretical predictions. More detailed experiments made outside Earth's atmosphere using the Cosmic Orbiting Background Explorer (COBE) satellite gave such good agreement that the audience applauded when results were shown at a meeting of the American Physical Society. Penzias and Wilson shared the Nobel Prize for their discovery. Penzias became a vice-president of AT&T so he probably does not have to clean pigeon droppings out of the antennas any more.

THERMAL Worksheet (Second Screen)

While viewing the Home screen of this worksheet press PgDn and you should see the second screen (Fig. 11-4) for blackbody radiation (Planck distribution). Figures 11-5 and 11-6 show results at 2.7K and 5K in two formats.

If your spreadsheet has the logarithmic scale capability you will find it useful to set graphs for this section as logarithmic on both the vertical and horizontal axes. Energy is expressed in electron volts (eV). Recall 1 eV = 1.6×10^{-19} J. The Planck distribution is given by,

$$E(f) = \frac{8\pi h}{c^3} \frac{f^3}{\exp(hf/kT) - 1} .$$

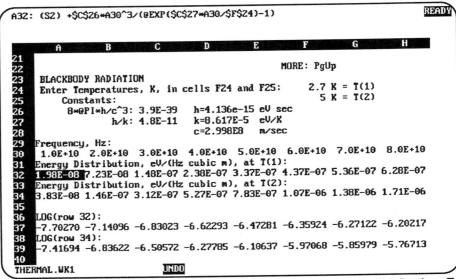

Fig. 11-4. Blackbody radiation energy density via the Planck distribution. Enter two temperatures in Cells F24 and F25. You may need to modify the frequency range because this distribution is very sensitive to temperature. Note that T(1) is 2.7 K, the present temperature of the cosmic radiation left from the Big Bang. Planck's constant *h* is in Cell D26, Boltzmann's constant *k* is in Cell D27, and the speed of light is in Cell D28.

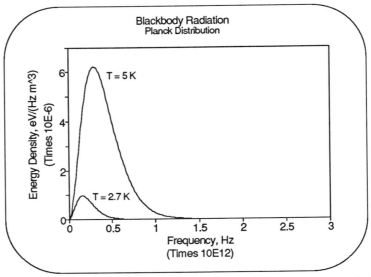

Fig. 11-5. Blackbody energy density at 2.7 and 5 K. Some materials become superconductors at these temperatures. Graph name: *Blackbody*.

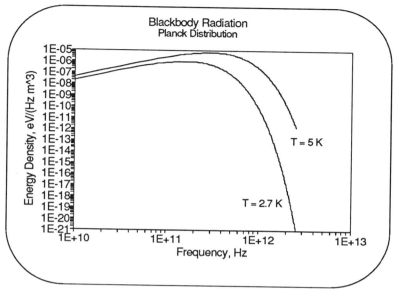

Fig. 11-6. This is the same information in Fig. 11-5, but in log-log representation.

Observe the shift of the maximum toward higher frequency as the temperature increases; this is called the Wien displacement law. The area under each curve is the total energy per cubic meter. The energy density increases as the fourth power of the absolute temperature; this is called the Stefan-Boltzmann law.

Equivalent Noise Bandwidth

This parameter (often called ENBW) is used to characterize data windows (see Chapter 9). It is the width of a rectangular window with the same peak power gain that would accumulate the same noise power as the window in use. Expressed in number of data bins, the rectangular is 1, von Hann 1.5, Hamming 1.36, Blackman-Harris 2, and Kaiser-Bessel 1.8. ENBW is tabulated for all common windows.

Quantization Noise

Additional noise, or what appears to be noise, on a digitized signal (not present on the original analog signal) comes from several sources:

- Code width (8 bit, 32 bit, etc.), which determines the theoretical best resolution

- Differential nonlinearity, a hardware error in the data acquisition system

- Multiplexing.

When an analog signal is digitized there is an inevitable error introduced because there is a smallest detectable change, or minimum resolution, of the analog-to-digital converter. The lack of an exact match appears as noise on the signal, and is called *quantization noise*. The smallest change represents 1 least significant bit (LSB) of the digitized value, and is usually called the *code width*. Increasing the number of bits increases the resolution. For example, an 8-bit A/D provides 256 steps and a 16-bit A/D provides 65,536 steps. A 16-bit A/D is acceptable for audio entertainment electronics.

The smallest detectable change in voltage is determined by the range, resolution, and gain of the data acquisition system. (Range, as used here, refers to the maximum and minimum voltages that the A/D converter can quantize.) As an example, consider a range of 0 to 10 V, a 16-bit A/D, and a gain of 100. For these parameters, the ideal code width or theoretical resolution of one bit in the digitized signal is

$$\frac{10V}{100 \times 2^{16}} = 1.5 \ \mu V$$

The digitized signal will appear to have microvolt-level noise even if the analog noise is less. Subsequent processing may increase the noise; an IIR filter recirculates noise through its feedback path. FIR filters do not have this problem.

SIGNOISE, Signal and Noise Worksheet

The Home screen, shown in Fig. 11-7, lets you experiment with signals and noise.

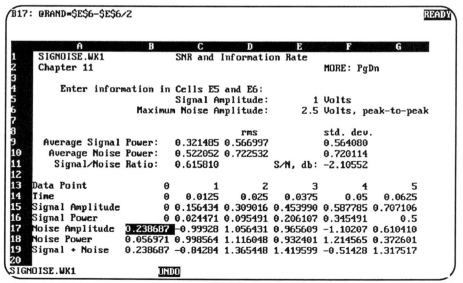

Fig. 11-7. Home screen of SIGNOISE. Cursor on cell B17 shows use of @RAND.

Worksheet Organization

Enter the signal amplitude in Cell E5 and the peak-to-peak noise amplitude in Cell E6.

Cell C9 shows the average signal power. This depends on the signal that you enter in row 15.

Cell D9 shows the rms signal.

Cell F9 shows thesignal standard deviation.

Cell C10 shows the average noise power.

Cell D10 shows the rms noise.

Cell F10 shows the noise standard deviation.

Cell C11 shows the signal/noise ratio.

Cell F11 shows the *SNR* in dB.

Row 14 is the Time axis. Use the Fill command to set the range and increment.

Row 15 is the noiseless signal. On the diskette at present there is a sinewave.

Row 16 is the signal power.

Row 17 is the user-defined noise. The formula is shown in Fig. 11-7. The noiseless signal and the noise are shown in Figs. 11-8 and 11-9.

Row 18 is the noise power.

Row 19 is the signal plus noise.

Press function key F9 to change to a new sample of noise. The noise distribution used here is "white" noise, indicating it has the same average power at all frequencies. You can pass this noise through a low-pass filter to obtain "pink" noise or a high-pass filter to obtain "blue" noise. Gaussian noise has a Gaussian *amplitude* distribution. Figure 11-10 shows a digital sound level meter.

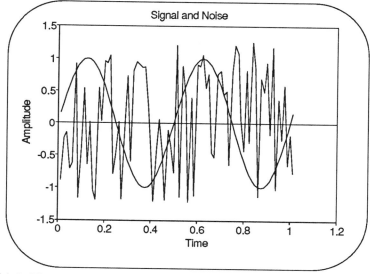

Fig. 11-8. The noiseless signal and the noise are shown before combining.

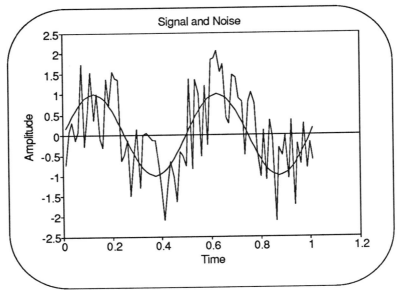

Fig. 11-9. Noiseless signal and the signal-plus-noise are shown superimposed.

Fig. 11-10. A sound level meter can be used to measure signal plus noise power. With the signal off the noise power can be measured independently. The instrument shown above that looks like a spacecraft is a digital sound level meter. It stores data and interfaces to a computer for downloading and additional processing. (Courtesy Brüel & Kjaer Instruments, Inc., Decatur, GA)

Information and Information Rate

The brain acquires information from various sensors, optical, auditory, olfactory, touch, and taste. Information is received by the brain as electrical signals which undergo elaborate processing. Computers are built along similar lines. In a quantitative sense Hartley defined information as,

$$Information = CBt \text{ bits}$$

where C is a constant depending on the system, B is the bandwidth in Hz, and t is the time interval in seconds during which the signal is transmitted.

Shannon showed that the maximum rate at which information can be transmitted or received is,

$$Information\ Rate = C \log_2(1 + S/N) \text{ bits/sec}$$

where S/N is the signal/noise power ratio. The $\log_2$ is for a binary system. In a spreadsheet you can use @IMLOG2 or recall,

$$\log_2(x) = \log_2(10) \log_{10}(x) \approx 3.3219 \log_{10}(x)$$

The maximum rate has never been achieved and it must be regarded as an ideal limit. It tells you the best you can expect. More complicated coding can increase the information rate beyond that of a simple binary system; the base of the log must be modified for other coding. Notice that at low S/N the information rate can be increased most effectively by increasing signal power. At high S/N the most effective way of increasing information rate is by increasing bandwidth. These things will become apparent when you use the second screen of SIGNOISE and its associated graphs.

SIGNOISE Worksheet (second screen)

While viewing the Home screen of SIGNOISE press the PgDn key and you should see something like Fig. 11-11. The second section lets you experiment with different bandwidths and signal/noise ratios to see their effects on the maximum information rate. You can set the signal and noise parameters independently; the signal and noise are then combined to produce a noisy signal. The rms values of each are shown, and the S/N ratio is calculated in ratio and dB format.

Enter three bandwidths in Cells C22, D22, and E22 as indicated.

Rows 26, 27, and 28 compute the information rate as a function of S/N.

Enter three S/N ratios in Cells C32, D32, and E32 as indicated.

Rows 35, 36, and 37 compute the information rate as a function of bandwidth.

Figures 11-12 and 11-13 show information rate as a function of S/N and bandwidth, respectively.

```
B25: (F0) +$C$22*3.3219*@LOG(1+B$24)                              READY
```

	A	B	C	D	E	F	G
1	SIGNOISE.WK1		SNR and Information Rate				
2	Chapter 11						
21		Enter Bandwidths (Hz) in Cells C22, D22, and E22:					
22			1000	5000	10000		
23							
24	S/N	0	0.5	1	1.5	2	2.5
25	Data rate, BW #1	0	585	1000	1322	1585	1807
26	Data rate, BW #2	0	2925	5000	6610	7925	9037
27	Data rate, BW #3	0	5849.575	9999.915	13219.16	15849.49	18073.39
28							
29							
30							
31		Enter S/N in Cells C32, D32, and E32:					
32			0.5	2	10		
33							
34	Bandwidth, Hz	0	5000	10000	15000	20000	25000
35	Data rate, S/N #1	0	2925	5850	8774	11699	14624
36	Data rate, S/N #2	0	7925	15849	23774	31699	39624
37	Data rate, S/N #3	0	17297	34594	51891	69188	86485
38							

```
SIGNOISE.WK1                  UNDO
```

Fig. 11-11. Second screen of SIGNOISE. The cursor on cell B26 shows the formula for the data rate (bits/s) at the first bandwidth.

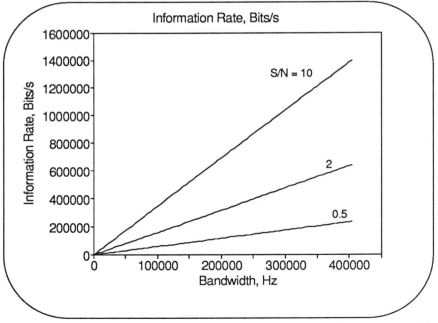

Fig. 11-12. Maximum information rate for a binary system for various *S/N*.

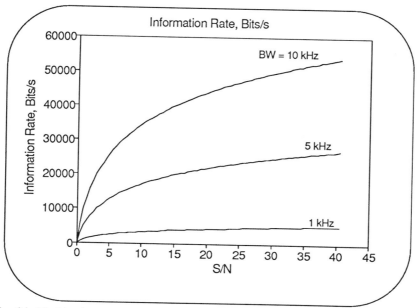

Fig. 11-13. Maximum information rate for a binary system for three bandwidths.

Noise, Total Harmonic Distortion, and Intermod

Noise and total harmonic distortion (THD) are associated because of the way they are measured. The input signal, usually cleaned up with a low-pass filter, is subtracted from the output and what is left is noise and THD. Harmonic distortion and intermodulation distortion, or "intermod," arise when the output of a system is not linearly proportional to the input. They are often referred to as nonlinear distortion or amplitude distortion.

Harmonic distortion occurs when Fourier components of the input produce harmonics. For the simplest case, a pure sine wave input at frequency f would produce new signals in the output with frequency $2f$, $3f$, $4f$, and so on. Depending on the system design either the odd harmonics or the even harmonics may dominate. The human ear has a logarithmic response so it produces strong harmonics that add richness to sound; that's why music sometimes sounds better when it's louder.

Intermod occurs when two or more signals mix due to nonlinearity. For example, two signals at frequencies f_1 and f_2 will mix to produce a series of new frequencies like $f_1 \pm f_2$, $2f_1 \pm f_2$, $f_1 \pm 2f_2$. In general, all mixtures of all harmonics will appear. This is a normal occurrence in the human ear but when additional intermod is produced in amplifiers the result is often undesirable. Signals used in making intermod measurements do not have to be spectrally pure because their harmonics will not usually cause significant intermod themselves and their harmonics will usually not be near the intermod products. It is important to distinguish intermod from *crosstalk*. Crosstalk or cross-coupling (also measured in dB) occurs when signals from one channel leak into another channel.

Two standard methods of measuring intermod, the SMPTE and the CCITT, are well accepted. The Society of Motion Picture and Television Engineers (SMPTE) uses two sine waves, one 50 times the frequency and ¼ the amplitude of the other. In a typical good stereo FM receiver the SMPTE intermod is 0.04%. THD is 0.025% at half power output and 0.04% at full power output.

The International Telephone and Telegraph Consultative Committee (Comité Consultatif International Télégraphique et Téléphonique) recommends using two closely spaced frequencies with equal amplitudes.

Radar Range Equations

This is a problem that people have worked on for many years, and it continues to provide a challenge for researchers. The maximum range at which a target can be detected depends on many things, but the *SNR* is one of the major factors. A simplified form of the radar range equation can be expressed as,

$$R_{max} = \left[\frac{P_t \, GA_e \, \sigma}{(4\pi)^2 S_{min}} \right]^{\frac{1}{4}}$$

where,

P_t is the transmitted power in watts

G is the antenna gain

A_e is the antenna effective aperture, m^2

σ is the radar cross section of the target, m^2

Entire books are written about σ and great effort has been expended to reduce the radar crossection of aircraft and ships.

S_{min} is the minimum detectable signal in watts. This is a critical parameter; it is much more desirable to decrease S_{min} than to increase transmitted power and antenna size.

If the receiver is not extremely low-noise then S_{min} may be approximated as,

$$S_{min} = kT_o \, B_n \, F \left(\frac{S_o}{N} \right)_{min}$$

T_o is assumed to be 290 K and B_n is the receiver noise bandwidth in Hz. Thermal noise power is given by,

$$n_o = kTB \text{ Watts} \quad \text{or} \quad n_o(dBm) = 10 \log_{10} \frac{kTB}{.001} \, .$$

Report 6930. "Part 2—Derivations of equations, bases of graphs, and additional explanations." Report 7010. These two reports are now collectors' items but copies can be obtained from many libraries, government document depositories, and technical databases.

Principles of Electronic Instrumentation, 3rd Ed., A. J. Diefenderfer (Saunders College Publishing, Philadelphia, 1993). ISBN 0-7216-3076-6. Chapter 13 is about noise and noise reduction in electronics. The presentation is at the undergraduate level and is understandable by people who aren't electrical engineers.

Radar Technology, E. Brookner (Artech House, Dedham, MA, 1977). ISBN 0-89006-021-5. The Appendix "How to look like a genius in detection without really trying" has a cookbook approach to target detection and radar range estimation, which is much more complicated than we have indicated.

Introduction to Random Signals and Applied Kalman Filtering, 2nd Ed., R. G. Brown and P. Y. C. Hwang (John Wiley & Sons, New York, 1992). ISBN 0-471-52573-1. The emphasis is on applications to the Global Positioning System but the early chapters are completely general.

A. Compagner, "Definitions of randomness," *American Journal of Physics* **59**, 700 -705 (1991). How do you know that a given sequence is random? How can you generate a true random sequence? These are continuing problems with many practical applications in such diverse fields as information theory, radar, computer science, cryptography, semiconductor physics, and the stock market. This paper discusses the issues and has 15 up-to-date references.

An Introduction to the Theory of Random Signals and Noise, W. B. Davenport, Jr., and W. L. Root (IEEE Press, New York, 1987). ISBN 0-87942-235-1. This is a republication of a book that has educated a generation of telecommunications engineers. The reader is assumed to have a high level of mathematical ability.

Signal Processing of Underwater Acoustic Waves, C. W. Horton, Sr. (Government Printing Office, Washington, D.C., 1969). Library of Congress Catalog Card Number 74-603409. The emphasis here is on sonar applications.

J. B. Johnson, "Thermal agitation of electricity in conductors," *Physical Review* **32**, 97–109 (1928). This is the original experimental work on thermal noise in conductors.

H. Nyquist, "Thermal agitation of electric charge in conductors," *Physical Review* **32**, 110–114 (1928). Nyquist's theoretical paper appeared back-to-back with Johnson's paper.

The Measurement of Time-Varying Phenomena, E. B. Magrab and D. S. Blomquist (Wiley-Interscience, New York, 1971). ISBN 0-471-56343-9. After a brief theoretical introduction the emphasis is on practical applications.

Symbols, Signals and Noise, J. R. Pierce (Harper & Row, New York, 1965). Library of Congress Catalog Card Number 61-10215. Pierce excels at explaining difficult concepts in a popular, understandable form.

Statistical Thermophysics, H. S. Robertson (Prentice-Hall, Englewood Cliffs, New Jersey, 1993). ISBN 0-13-845603-8. Chapter 10 discusses current ideas of noise and fluctuations.

Introduction to Radar Systems, M. I. Skolnik (McGraw-Hill, New York, 1962). Library of Congress Catalog Card Number 61-17675. Skolnik's book is a golden oldie that covers the fundamentals in a clear and comprehensive style. Chapter 2 discusses the radar range equation and Chapter 9 presents the fundamentals of detection of signals in noise. Chapter 10 is devoted to extraction of information from radar signals.

C. E. Shannon, "The mathematical theory of communication," *Bell System Technical Journal* **27**, 379–424, 623–657 (1948). This is Shannon's original paper that started information theory and modern telecommunications.

Principles of Underwater Sound, 2nd Ed., R. J. Urick (McGraw-Hill, New York, 1975). ISBN 0-07-066086-7. The sonar range equation shares many of the basic principles of the radar equation but the problems of sonar propagation and detection are profoundly affected by the medium, boundaries and noise sources. In some ways it makes the radar problem seem easy. Sonar has one major difference: Passive listeners can detect and classify targets.

R. W. Wilson, "The cosmic microwave background radiation," *Reviews of Modern Physics* **51**, 433–445 (1979). This is the lecture delivered by Wilson on the occasion of his acceptance of the Nobel Prize. It contains a detailed discussion of their experiments and sources of noise. The lecture by Penzias immediately precedes this. For recent experimental results with the COBE satellite see *Physics Today*, March 1990, p. 17.

Noise in Solid State Devices and Circuits, A. Van der Ziel (John Wiley & Sons, New York, 1986). ISBN 0-471-83234-0. This is a classic work.

What amazed me most in my first calculations on this effect was the magnitude of the signals which one could expect from nuclear induction. In our example of a cubic centimeter of water in a normal field of a few thousand gauss they turned out to amount to the order of a millivolt. This magnitude is well above the noise which accompanies any radio receiver and which sets the ultimate limit of signal detection.

Felix Bloch, "The principle of nuclear induction"
Nobel Lecture, December 11, 1952.
(*Nobel Lectures,Physics,* Vol. 3, 1942-1962, Elsevier, Amsterdam, 1964)

Appendix 1

Decibel Definitions

Common Definitions

dBc

c = Carrier. The reference is the power of the carrier wave. The symbol dBc is used in telecommunications, for example, to measure noise on a signal.

dB(HL)

HL = Hearing Level. The reference is the normal standard for threshold at various frequencies. This must be specified as either ASA 1951 Standard or ISO 1964 Standard.

dB(HTL)

HTL = Hearing Test Level. This is the same as dB(HL), and is a variant used by some authors.

dB(IL)

IL = Intensity Level. This is the same as dB(SIL).

dBm

m = milliwatt. The reference power is one milliwatt = 10^{-3}W.

dB(SIL)

SIL = Sound Intensity Level. This usually means that the reference is 10^{-12} W/m^2.

dB(SPL)

SPL = Sound Pressure Level. The standard definition has the reference equal to 2×10^{-5} N/m^2. (One Newton/meter^2 = one Pascal = one Pa.)

dBV

V = Volt. This indicates that dB is measured as a voltage ratio. The reference voltage should be specified, but is usually one volt. Both voltages must be measured across the same value of resistance or impedance!

dBW

W = Watt. This indicates that dB is measured as a power ratio. The reference power is one Watt. This is sometimes written as dBw.

Specialized Definitions

In the following, TLP means Test Level Point; CCITT refers to the International Telegraph and Telephone Consultative Committee, and CCIR refers to the International Radio Consultative Committee. CCITT and CCIR operate under the ITU, the International Telecommunications Union, Geneva, Switzerland.

dBA or dB(A)

An acoustic term based upon the 40.0-phon equal loudness contour. The phon is a unit of perceived loudness and equals the sound intensity of a 1000.0 Hz tone referred to 10^{-12} W/m^2 as judged by listeners to match the loudness at other audio frequencies. See Chapter 2, page 28.

dBa

This indicates dB above adjusted reference noise (–85 dBm). This is an F1A weighted term formerly employed by the U.S. telephone industry to indicate channel noise. This term, now obsolete, is being replaced by the dBrn. The dBa is based on a –85.0 dBm annoyance perceptibility threshold. The F1A weight is used in the Western Electric 2B Noise Measuring Set.

dBa0

This is related to the dBa, and indicates the random noise intensity at the 0.0 dBm TLP.

dB(B)

This is the same as dB(A) but is based on the 70.0 phon equal loudness contour. See Chapter 2, page 28.

dB(C)

This is the same as dB(A) but is based on the 100.0 phon equal loudness contour. See Chapter 2, page 28.

dBc

This is sometimes called "dB Collins" in which 0.0 dB is 0.775 V across any impedance. It should not be confused with "dB carrier" which has the same abbreviation. The context in which dBc is used should be helpful in preventing confusion.

dB/c

This denotes attenuation per 100.0 feet. This is used, for example, in coaxial cable, waveguide, or fiber optic cable measurements.

dB(D)

This refers to an acoustical equal noisiness contour.

dBd

This is used to indicate antenna gain relative to that of a half-wave dipole antenna.

dBe

This is used by Siemens GmbH, referenced to 0.775 V across a low impedance.

dBeff

This is used by Rhode & Schwarz GmbH, referenced to 0.7 V rms.

dBer

This is also used by Siemens, and is the same as dBe.

dBi

This indicates antenna gain referenced to an isotropic radiator.

dBj

This is used by Jerrold Electronics, Inc., for radio signals, referenced to 1,000 microvolts.

dBk

The reference is 1.0 kW.

dB/m

This indicates dB attenuation per meter; used to characterize coaxial cable, fiber optic cable, waveguide, and bulk material.

dBma

This is referenced to a power level of 1.0 milliwatt, F1A weighted (Western Electric 2B Noise Measuring Set).

dBmc

This is the same as dBma, but C-message weighted. (See dBrn.)

dBm0

This indicates the power level in dBm referred to the 0.0 dBm TLP. This term is sometimes employed to indicate dB relative to a given TLP. For example, a +12.8 dBm0 white noise loading at a −33.0 dBm TLP would denote a white noise loading power level of −20.2 dBm.

dBm0p

This is based on 1.0 milliwatt at the 0.0 dBM TLP, psophometrically weighted.

dBm0s

This is a CCITT term, referenced to "zero" program level (s = sound for TV-audio or radio broadcasting), as opposed to telephone communications.

dBm0t

This is a CCITT term, referenced to "zero" telephone level, as opposed to sound TV-audio or radio broadcasting.

dBp

This is used to indicate signal-to-noise ratio, psophometrically weighted.

dBpp

This is used by Rhode & Schwarz, primarily in TV, referenced to 0.7 V peak-to-peak.

dBr

This means "decibels relative." For example, a +12.8 dBr noise loading at a −33.0 dBm0 TLP would yield a −20.2 dBm noise power.

dBrn

This indicates "decibels above reference noise." It is based on 0.0 dBm = −90.0 dBm as employed in the Western Electric 2B Noise Measuring Set (144-weighting), and the Western Electric 3A (C-message weighting).

dBr0

This is a 144-weighting term (Western Electric 2B Noise Measuring Set), based on a –90.0 dBm power level at the 0.0 dBm TLP.

dBrnc

This is referenced to noise, C-message weighted (0.0 dBrnc = –90.0 dBm).

dBrnc0

This is a C-message weighted value (Western Electric 3A Noise Measuring Set) based on –90.0 dBm measured at the 0.0 dBm TLP.

dBrap

This indicates "decibels referenced acoustic power." It is referenced to –160.0 dBW.

dBrms

This is a Rhode & Schwarz "TV decibel," referenced to 0.7 V rms.

dBss

This is the same as dBpp.

dBT

This is a sound pressure level referenced to the noise in the driver's position of an average 1965 T-bird, windows down, moving at 55 miles per hour on a level, dry asphalt surface.

dBu

This is used by Siemens, referenced to 0.775 V. Context must be used to decide between this or the next dBu, which is actually dBμ.

dBu

This is actually dBμ or dB-micro, referenced to 1.0 microvolt. Context must be used to decide between this or the Siemens dBu.

dBU

This means "decibels unicorn." See Chapters 2 and 3.

dBv

dBv is a variant of dBV, indicating the reference is 1.0 V.

dBvg

This means "decibels voltage gain."

dBx

This means "decibels reference coupling," a cross-coupling term.

PNdB

This means "decibels based on perceived noise."

Db as a function of power ratio and voltage ratio is shown in Fig. A1-1. Figure A1-2 shows dBm as a function of voltage (mV) for three common transmission line impedances.

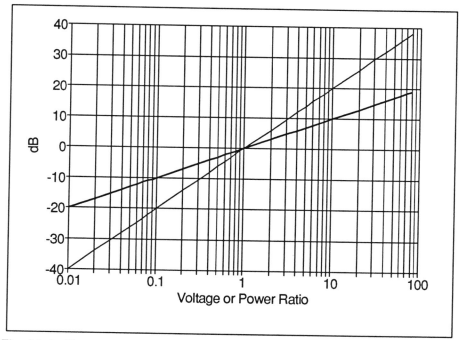

Fig. A1-1. dB as a function of power ratio (thick line) and voltage ratio (thin line).

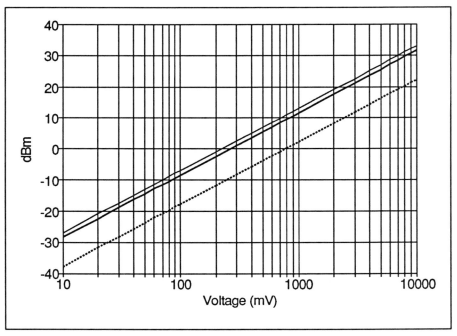

Fig. A1-2. dBm as a function of voltage for 50Ω (thin line), 70Ω (thick line), and 600 Ω (dotted line).

Appendix 2

Signal Parameters

RMS

We need to discuss how intensity and pressure enter into our expressions for decibels. This is necessary so that you will understand why we measure dB(SPL) using rms pressure and dB(SIL) using rms intensity. It will also become clear why we use rms volts or rms power instead of instantaneous volts and power.

Sound waves are disturbances that travel in space and time, so they can be represented as functions of space and time. These disturbances consist of small changes in pressure above and below the ambient atmospheric pressure. These changes are very small compared to the prevailing pressure.

How small are they? They are very, very, small. Atmospheric pressure is about 10^5 N/m^2. The faintest sound detectable by the human ear, corresponding to 10^{-12} W/m^2, involves a pressure change of 2.9×10^{-5} N/m^2. Thus, the faintest sound involves a change of about 3 parts in 10^{10}. This is astounding when you first learn about it, and becomes even more astounding as time passes. Sound consists of a tiny disturbance, about 3 parts in 10^{10}, riding on a very large constant atmospheric pressure.

Speech and music are particularly complicated because they are partially predictable and partially unpredictable. We can distinguish two broad categories of signals as periodic (predictable) and random (unpredictable). You can produce a predictable signal by humming a sound that doesn't change in pitch or loudness. You can produce an unpredictable signal by whispering the sound corresponding to "Shhhh" or letting the air out of a bicycle tire. Even the periodic part of sound is predictable only over a short time interval; no one can say with certainty what the next word, phoneme, or musical note will be.

They say you can't compare apples and oranges, but that is exactly what we need to do. In order to measure intensities and sound pressures of complicated waves a common denominator is needed. That common denominator is the average power.

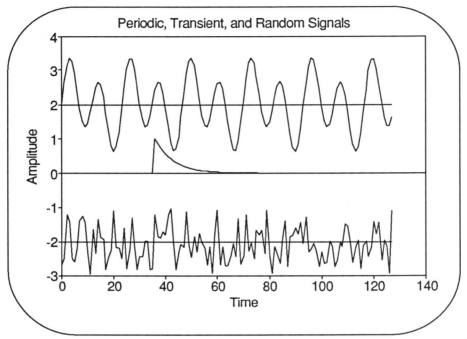

Fig. A2-1. Examples of signal types. Upper: Two sine waves combine to produce a periodic signal. Middle: A transient, nonperiodic signal. Lower: A random, nonperiodic signal. Signals often consist of a combination of all three types.

It can be proved (trust me) that the power carried by a wave, no matter how complicated, is proportional to the mean square of the amplitude of the wave. Consequently, the square root of the mean square of the amplitude is a common denominator in comparing waves of different shapes.

The amplitude of a sine wave or any other wave is constantly changing, so the instantaneous power is constantly changing. It would be fruitless to try to compare instantaneous powers. The time-averaged power and its correlate, the rms value of the amplitude, are useful in comparing these apples and oranges.

Since intensity is average power per unit area (W/m^2) we use the *average* intensity when we use intensities to calculate decibels. Also, since power is proportional to the square of the pressure we use the square root of the mean square of pressure, or root-mean-square pressure, or rms pressure when we calculate in decibels.

It is common practice to take all intensities as time-averaged intensities and all pressures as rms pressures, where the average is taken over a long time interval. Similarly, voltages are often expressed as their rms values unless otherwise noted (see the example later concerning the rms and maximum voltage at electrical outlets). Most inexpensive voltmeters are calibrated to read rms values for a sine wave and will give erroneous readings if other waveforms are used. An analog true rms voltmeter will measure, by thermal means, the rms value of any waveform, including a completely random wave such as noise. Because of their method of measuring, analog true rms voltmeters have rather slow response times. Digital instruments can be designed to compute the rms value rapidly. Conside the example in Table A2-1. (Pressure is in Pa, taken every 0.1 millisecond)

Table A2-1. rms Example	
Pressure Data	Square
1	1
−2	4
5	25
−4	16
Total: 0	Total: 46
Mean: 0	Mean Square: 46/4 = 11.4
	rms: $(11.5)^{1/2}$ = 3.39

In this example the average value of the pressure is zero, but the rms value is 3.39. So, to find the rms value, find the square root of the mean square. The rms values of some common signals are shown in Table A2-2.

Table A2-2. rms Values of Common Periodic Waveforms (peak value = 1)	
Sine	$0.707 = 1/2^{1/2}$
Full-wave rectified sine	0.707
Half-wave rectified sine	0.500
Square	1.000
Triangle	$0.816 = 2/(3^{1/2})$
Sawtooth	$0.577 = 3/(3^{1/2})$

Of all the types of signals, the sine wave has a special place in our hearts, minds, and ears. It is beautiful. It is a pure tone, a single frequency. It has many strange and wonderful properties. There is a continuing worldwide contest going on to see who can make the purest sine wave; at least two Nobel prizes (the laser and the Mössbauer effect) have been won so far for this. From a Fourier perspective, the sine wave is the fundamental building block of nature and it is periodic and predictable, and, if you're not so picky about spectral purity, it is easy to generate. In many respects it is an ideal test signal because if we use it as input to a system we can measure the amplitude and phase of the system response (output) in a particularly simple and easily interpretable way. For a sine wave only, the rms value is 0.707 times the amplitude (recall that 0.707 is half the square root of 2).

Sine wave only:

$$\text{rms value} = \text{amplitude}/2^{1/2} = \text{amplitude}/1.414 = 0.707 \times \text{amplitude}$$

Example:
Pure Tone Amplitude and rms

Measured with a calibrated microphone and oscilloscope, suppose a pure tone (sinewave) has an amplitude of 2×10^{-10} N/m^2. The rms value is therefore,

$$(0.707)(2 \times 10^{-10}) = 1.414 \times 10^{-10} \text{ N/m}^2.$$

Example
Voltage at Home Power Outlet

The rms value of the voltage commonly used in home alternating current circuits is 120 volts. Because the alternating current and voltage are sine waves their average values are zero. The peak value of the voltage is the rms value multiplied by $2^{1/2}$, or $120 \times 1.414 = 169.7$ volts. Thus the maximum voltage available at a 120 V receptacle is almost 170 V.

Average value = zero

rms value = 120.0 V

Peak value = 169.7 V

Other Useful Signal Parameters

Amplitude Probability Distribution

Still other useful parameters are related to signal probability distributions. One of these is the amplitude probability distribution. The amplitude probability distribution answers a question of the type, What is the probability that I will measure a sound pressure between p_1 and p_2? The intensity probability distribution is similar, and answers a question of the type, What is the probability that I will measure an intensity between I_1 and I_2?

Aspect Ratio

The aspect ratio which is the ratio of the number of peaks to the number of zero-crossings in a given time; some signals will fall and then rise many times without crossing zero. The aspect ratio is a measure of this wiggliness and if the aspect ratio is large it indicates a high harmonic content or a broad spectrum (which is the same thing).

Auto-Correlation Function

This is a measure of how a function correlates or compares with itself. See Chapter 10.

Cepstrum

This is the inverse Fourier transform of the log of the spectrum magnitude. The cepstrum is an aid in separating signals that have been convolved; it is often used to identify and separate echos in multi-path propagation. The log of the complex spectrum is used in complex cepstrum analysis. See Chapter 10.

Crest Factor

The crest factor, the ratio of the peak value to the rms value, is an important parameter which describes this peakedness. For example, from the above data for the sine wave the crest factor is $169.7/120 = 1.414$. For an extreme example, shooting a pistol every ten minutes in an otherwise quiet room would produce a sound signal with a very large crest factor.

Cumulative Probability Distribution

The cumulative probability distribution (one for amplitude and one for intensity) answers the question, What is the probability that I will measure a value of at least p_1 or I_1?

JTFA (Joint Time-Frequency Analysis)

The waveform tells you when something is happening but it doesn't tell you what frequencies are involved. The spectrum tells you what frequencies are present but it doesn't tell you when they are present. JTFA addresses this problem by means of a three-dimensional plot of the signal. The x-axis is usually time and the y-axis is usually frequency. The z-axis can be represented by different colors for different amplitudes or power, or the plot can be a "waterfall" graph in a pseudo-3D presentation.

Mean (Average) Value

The mean value of the sequence $[x_n]$ is defined as,

$$m_x = \frac{1}{N} \sum_{n=0}^{N-1} x_n$$

It is often useful to subtract the mean value from a data set before performing the FFT operation. This removes the spike at zero frequency. The spreadsheet function AVG is used to compute the mean value in the spreadsheet.

Parseval's Theorem

This theorem states the almost obvious fact that the total energy in a waveform is the same as the total energy in its spectrum (see Spectrum). This means that the sum of the squares of all the data in a waveform is equal to the square of the magnitude of all of the data in the spectrum, divided by the number of data points. You can see this in all of the worksheets involving the FFT. Because of Parseval's theorem the spectrum power is usually normalized by the number of data points. This theorem is one way to check the accuracy of a spectrum. Precisely, Parseval's theorem for discrete data tells us,

$$\sum_{n=0}^{N-1} |s_n(t)|^2 = \frac{1}{N} \sum_{n=0}^{N-1} |S_n(f)|^2 \ .$$

Peak Value

The peak value is very important in situations where the signal is of high intensity for a short time; the peak value may be very large even though the rms and average values are small. MAX detects the peak value in a row or column of data in the spreadsheet.

Peak-to-Peak Value

Use MIN together with MAX to compute the peak-to-peak value. The spreadsheet operation is MAX(..) − MIN(..).

Power Spectral Density

The PSD, a quantitative measure of the signal power at each frequency, is a real function of frequency. The cross-PSD is a measure of the common power of two signals at every frequency. the PSD and cross-PSD comprise Fourier transform pairs with the autocorrelation function and the cross-correlation function, respectively. See Chapter 8.

Schlangeheit

This parameter is related to aspect ratio; it is the time average of the absolute value of the second derivative of the waveform. Schlangeheit, like aspect ratio, is a number that gives a measure of the frequency content of a signal without actually computing the spectrum.

Spectrum

The spectrum is a quantitative measure of the signal components at each frequency. The spectrum is generally complex, and can be represented as its magnitude and phase, or equivalently as its real and imaginary parts. See Chapter 8.

The Discrete Fourier Transform (DFT) can be defined in several ways. Bracewell's representation has been widely accepted in engineering and scientific works,

$$S(k) = \frac{1}{N} \sum_{\tau=1}^{N} s(\tau) \exp[-j2\pi(k/N)\tau]$$

(forward)

and

$$s(\tau) = \sum_{k=1}^{N} S(k) \exp[j2\pi(\tau/N)k] \ .$$

(inverse)

Here the negative exponential appears in the forward transform and the positive exponential appears in the inverse transform. The normalizing factor N appears in front of the forward transform. In some works on mathematics the signs in the exponentials are reversed, and the normalizing factor $\sqrt{N}$ appears in front of both the forward and inverse transforms. The FFT and IFFT operations in the spreadsheet speed up computation of the transforms by taking advantage of redundancy and symmetry.

Standard Deviation

The standard deviation σ is the square root of the variance. STD and STDS perform this computation in spreadsheets.

Variance

Variance is a common measure of the spread of a data set about its mean value m_x. The variance is the square of the standard deviation σ. The *sample variance* (spreadsheet VARS) is defined as,

$$\sigma^2 = \frac{1}{N-1} \sum_{n=0}^{N-1} (x_n - m_x)^2$$

The *population variance* (spreadsheet VAR) is defined as,

$$\sigma^2 = \frac{1}{N} \sum_{n=0}^{N-1} (x_n - m_x)^2$$

For an example see the worksheet CONVOLVE in Chapter 9.

If the mean value m_x is zero and N is much greater than 1 the variance is approximately equal to the mean square and the standard deviation is approximately the rms value. When N is large the population variance and sample variance are almost equal.

Fig. A2-2. This sound level meter is being used to measure periodic, transient, and random sound near a highway. (Courtesy of Brüel & Kjaer Instruments, Inc., Decatur, GA)

Appendix 3

Hearing Aid Tests

What this Appendix is About

Hearing aid tests, bandwidth, and gain differ considerably from standard engineering measurements. The system is influenced by the acoustic properties of the input and output transducers and couplers as well as by the electronics. Nonlinearities of automatic gain control (AGC) and dynamic amplitude compression further complicate the problem. This is an area of continuing research and development.

Definitions

Saturation Sound Pressure Level for 90 dB Input Sound Pressure Level

This is the sound pressure level developed in a 2 cm^3 coupler when the input sound pressure level at the microphone opening of the hearing aid is 90 dB(SPL) with the gain control of the hearing aid full-on. The abbreviation for this term is SSPL90.

High-Frequency-Average Saturation Sound Pressure Level

This is the average of 1000, 1600, and 2500 Hz values of SSPL90. The abbreviation for this term is HF-average SSPL90.

High-Frequency-Average Full-On Gain

This is the average of 1000, 1600, and 2500 Hz values at full-on gain. The abbreviation for this term is HF-average full-on gain.

Reference Test Gain Control Position

This defines the setting of the hearing aid gain control so that the average of the earphone coupler sound pressure levels at 1000, 1600, and 2500 Hz, with a pure tone input sound pressure level of 60 dB, is 17 dB less than the HF-average SSPL90.

If the gain available will not permit this, or if instrument is an AGC (Automatic Gain Control) aid, the full-on gain control position of the hearing aid is to be used.

"Basic Setting" of Controls

Tone control settings should be chosen to give widest frequency range. All other control settings should be chosen to give the greatest HF-average SSPL90 and the highest HF-average full-on gain. If this is not possible, the setting giving the greatest HF-average SSPL90 should be selected. If varying degrees of AGC are available, tests are to be made with greatest amount of compression, even if a reduced HF-average SSPL90 results.

Recommended Measurements

SSPL90 Curve

With the gain control full-on and with basic settings of controls, record or otherwise develop a curve of coupler sound pressure level versus frequency over the range 200 - 5000 Hz, using a constant input sound pressure level of 90 dB.

HF-Average SSPL90

Measure and take the average of the 1000, 1600 and 2500 Hz SSPL90 values.

Full-On Gain Curve

Full-on gain is measured with the gain control set to its full-on position and with a sinusiodal input sound pressure level of 60 dB or, if necessary to maintain linear input-output conditions, with an input sound pressure level of 50 dB. For AGC aids, the input sound pressure level must be 50 dB.

HF-Average Full-On Gain

Measure and take the average of the 1000, 1600 and 2500 Hz full-on gain values.

Frequency Response Curve

With the gain control in the reference test position, and with an input sound pressure level of 60 dB, record the frequency response curve over the range 200 - 5000 Hz or a lesser range determined by limits 20 dB below the average of the 1000-, 1600-, and 2500-Hz response levels. For AGC aids, an input sound pressure level of 50 dB is to be employed.

Harmonic Distortion

With the gain control in the reference test position and with an input sound pressure level of 70 dB, measure and record the total harmonic distortion in the coupler output for input frequencies of 500, 800, and 1600 Hz. (If the response curve rises 12 dB or more between any distortion test frequency and its second harmonic, distortion tests at that test frequency may be omitted.)

Equivalent Input Noise Level (L_n)

With the gain control in the reference test position, determine the average of the coupler sound pressure levels at 1000, 1600, and 2500 Hz for an input sound pressure level of 60 dB (L_{av}). Remove the acoustic signal and record the sound pressure level in the coupler caused by the inherent noise of the hearing aid (L_2). Then the equivalent noise is calculated by $L_n = L_2 - (L_{av} - 60)$ dB. *Note:* This method is not appropriate for AGC aids.

Battery Current

With the gain control in the reference test position, measure the battery current with a pure-tone 1000-Hz input signal at a sound pressure level of 65 dB.

Coupler Sound Pressure Level with Induction (Telephone) Coil

With the gain control full-on and the hearing aid set to the "T" (telephone input) mode, the hearing aid is placed in an alternating magnetic field having a frequency of 1000 Hz and a magnetic field strength of 10 mA/m, and oriented to produce the greatest coupler sound pressure level.

Additional Recommended Measurements for AGC Aids

The following additional tests apply to AGC hearing aids, which are defined as hearing aids incorporating means (other than peak clipping) by which the gain is automatically controlled as a function of the magnitude of the input signal. The full-on gain control position of the hearing aid is to be used.

Input-output characteristics

Using a pure-tone test frequency of 2000 Hz, measure and plot the coupler sound pressure level for input sound pressure levels from 50 to 90 dB, in 10 dB steps.

Dynamic AGC Characteristics

Using a square wave modulated pure tone input signal of 2000 Hz, determine the attack time defined as the time between an abrupt input level increase from 55 to 80 dB and the point at which the output has stabilized to within 2 dB of the steady-state value for the 80-dB input. The release time is defined as the interval between the abrupt input level drop from 80 to 55 dB and the point at which the output has stabilized to within 2 dB of the steady-state value for the 55 dB input.

Table A3-1. Summary of Typical Specifications and Tolerances	
Specification	*Tolerance*
1. SSPL 90 curve	Maximum value of SSPL 90 shall not exceed specified value
2. HF-Av. SSPL 90	Must be within 4 dB of specified value.
3. Full-on Gain curve	No tolerances.
4. HF-Av. Full-on Gain	Must be within 5 dB of specified value.
5. Frequency Response Curve (see Note A below)	(See Note B below)
6. Harmonic distortion (see Note A below)	Must not exceed specified values at 500, 800, and 1600 Hz.
7. Equivalent Noise Input (see Note A below)	Must not exceed specified maximum value
8. Battery current (see Note A below)	Must not exceed specified maximum value.
9. Induction coil (1000 Hz)	Must be within 6 dB of specified value.
10. AGC Input-Output Characteristic	Match measured and specified curves at 70-dB level. Measured values at 50 and 90 dB must be within 4 dB of specified values.
11. AGC Attack or Release Times	Must be within 5 milliseconds or 50% of specified values, whichever is larger.

Note A: Reference test gain control position is used in Items 5, 6, 7, and 8. Gain control must be set 17 dB (+1 dB) below HF-average SSPL90 for each individual instrument.

Note B: Frequency Response Curve
   a. From the manufacturer's specified frequency response curve, determine the average of the 1000-, 1600-, and 2500-Hz response levels.
   b. Subtract 20 dB to obtain the "reduced level."
   c. Draw a straight line parallel to the abscissa at the reduced level.
   d. Note the lowest frequency, f_1, at which the frequency response curve intersects the straight line.
   e. Note the highest frequency, f_2, at which the frequency response curve intersects the straight line.

Frequency Range: For information purposes, but not for tolerance purposes, the frequency range (bandwidth) of the hearing aid is considered as being between f_1 and f_2.

The tolerances in the two bands are often defined as follows:

Table A3-2. Frequency Response Tolerance		
Band	Frequency Limits	Tolerance
Low Band	1.25f_2 to 2000 Hz	4 dB
High Band	2000 to 4000 Hz or 0.8f_2, whichever is greater.	6 dB

Compliance with these specifications may be determined using a template with upper and lower limits derived from the manufacturer's specified response curve. Unlimited vertical adjustment of the template is permitted. Horizontal adjustment up to 10% in frequency is permitted. Following these vertical and horizontal adjustments, the entire low and high band measured curve must lie between the upper and lower limit curves.

Digging Deeper

Instrumentation, An Introduction for Students in the Speech and Hearing Sciences, T. N. Decker (Longman, White Plains, New York, 1990). ISBN 0-8013-0152-1.

Complete current standards may be obtained from:
American National Standards Institute (ANSI), 11 West 42nd Street, New York, NY 10036. Phone (212) 642-4900.
ANSI (S3.22-1987) Specification of hearing aid characteristics.
ANSI (S3.35-1985b) Methods of measurement of hearing aids under simulated *in situ* working conditions.

"Testing electroacoustic performance of ASP and nonlinear hearing aids," J. Heide, *Hearing Journal* vol. **41**, No. 4, 33 - 35 (1987). This paper discusses compression and signal processing aids.

Appendix 4

Inscrutable Wonders
of the Universe

There are 57 Inscrutable Wonders of the Universe, commonly abbreviated IWU. Some of the more familiar IWUs are *The True Meaning of Life*, *The Cosmic Forces That Shape Our Destiny*, and so on. I regret that space does not permit giving a complete explanation of all of them here. In this appendix we will only discuss some of the more important IWUs.

IWU #7. The Tao of the Transcendentals

Once upon a time, long ago and maybe far away, in the dim dark days beyond recall, our ancestors started counting. Some people say ancestors don't count; those people are wrong.

The word transcendental has a meaning in the real world that is different from its meaning in philosophy. Transcendental is a big word that was applied when it was thought that transcendental numbers were more rare than herds of unicorns, so people would not have to spell the word very often. These numbers are not the roots of any algebraic equation with integer coefficients and we will never know all about them because they appear to us as never-repeating unending decimals, weird continued fractions, and bizzare but simple infinite series.

Well, along came Georg Cantor, who was born in 1845 in St. Petersburg (Russia, not Florida). While he was looking into the realm of the infinite and beyond he showed that of all the numbers, the transcendental ones are the most common, that is, they are the least transcendental. *C'est la vie.*

The first transcendental number to be discovered was π (pronounced "pie," as in "Easy as π"). It was found thousands of years ago and was known to be greater than 3 and less than 4 but people didn't know it was transcendental. Archimedes, who was killed in 212 B.C. by a stupid Roman soldier who had orders not to kill him, knew that π was more than 223/71 and less than 22/7. The Bible is less accurate on the subject of transcendental numbers; in Chronicles and The Book of Kings the value of π is given as 3. That is strange, because Egyptian mathematicians knew that π was approximately 3.16. Perhaps there was a problem of communication.

By 150 A.D. Ptolemy knew that π was approximately 3.1416. Of course, they didn't write the numbers like this because the decimal point, positional notation, and Arabic numerals hadn't been invented. Checkbooks hadn't been invented either because nobody could figure out how to use Roman numerals except for numbering chapters in books and, later, for movie copyrights.

One of our other favorite transcendental numbers, the base of natural logarithms e, was found much later than π, and the eminent French mathematician Hermite proved in 1873 that e is transcendental. Only later, in 1882, did Lindemann prove that π is transcendental. Nobody knows yet whether γ, Euler's constant, is transcendental or not; the suspense is dreadful. (*Note*: $\gamma = 0.5772156649\ldots$)

The number e is unique in many ways. One of its extraordinary properties appears in calculus. Consider the basic integral,

$$\int x^n \, dx = \frac{x^{n+1}}{n+1} + \text{a constant.}$$

This is true for any n, integer or not, except for $n = -1$. It is obvious that the result blows up at $n = -1$. But, for $n = -1$ only,

$$\int x^{-1} dx = \ln(x) + \text{a constant.}$$

Another truly remarkable property is that e^{kx} is invariant to integration and differentiation (here k is a constant). We know that integration tends to smooth functions and differentiation tends to accentuate changes. Here we see that e^{kx} is in a sense the smoothest of all functions because it is invariant to integration and differentiation. Integration does not make it smoother and differentiation does not make it rougher.

$$\int e^{kx} \, dx = \frac{e^{kx}}{k} + \text{a constant}$$

$$\frac{d}{dx} e^{kx} = k \, e^{kx}$$

So, integration and differentiation merely change e^{kx} by a scale factor, leaving the function unchanged. This is the *only* function with this property. Because of this, ordinary sines and cosines also have this unique property (see IWU #30, below).

You want a couple of hot numbers? Try $2.718281828\ldots$ and $3.141592654\ldots$

IWU #17. Imaginary Numbers

You do not have to believe anything in the following discussion, but it only works if you believe.

As if transcendental numbers weren't enough to keep us occupied for a few thousand years, along came the imaginary numbers. Perhaps this was a poor choice for a name. Well, what's in a name? Even though we may not understand them we use them 24 hours a day everywhere in the world and on spacecraft that we launch toward other planets. We use imaginary numbers to design the instruments to detect light from the most distant stars and galaxies, and we use imaginary numbers to interpret these signals. We use them because we need them and they simplify things. We use a lot of other things for less cogent reasons.

Imaginary numbers were the subject of heated arguments and many reputable mathematicians resisted their intrusion. People got angry. Confrontations occurred. It is not a pretty sight when mathematicians become violent. They had reluctantly accepted the zero, although it was nothing. Negative numbers were more difficult to accept; how could you have less than nothing? These were real problems in the conceptual development of human thought.

The proponents of imaginary numbers prevailed; could anyone say that the equation $x^2 + 1 = 0$ has no solution? The solution is, of course, $x = \pm j$, if we define $j^2 = -1$. (Often you will find the imaginary unit written as "i" by mathematicians and "j" by scientists, engineers, and some ordinary mortals. Some of these people use the symbols without question. Some use the symbols but still don't believe; to them the Gates of Heaven shall be forever closed.) Figures A4-1 shows one use of j.

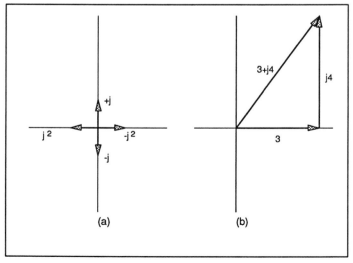

Fig. A4-1. (a) The symbol j is a signpost that only tells four directions. (b) Combined with real numbers it can represent a vector.

If you don't worry about immortality and you just want to have fun you can consider j to be like a sign-post that only tells direction and not how far. In fact, as a sign-post it's rather restricted because it will only point in two directions, $+j$ and $-j$ along the imaginary axis. You can get two more directions by multiplying $(j)(j)$ and $(j)(-j)$ which point along the real axis to -1 and $+1$. By combining j with real numbers you can make good sign-posts that will point in any direction and tell how far to go.

The useful graphic description shown in Fig. A4-1 was discovered by a Norwegian surveyor named Wessel and a French bookkeeper named Argand. It is usually called an Argand diagram; *sic transit gloria*, Herr Wessel.

The vector in the figure can be written as $3 + j4$. This is called a *complex number* because it is a mixture of real and imaginary. You cannot count imaginary numbers on your fingers and toes. In order to find the magnitude of a complex number we multiply the number by its complex conjugate and take the square root; to make a number into its complex conjugate you simply change the sign of the imaginary part.

It goes like this:

$$\text{Square of magnitude} = (3 + j4)(3 - j4)$$

$$= 9 - j12 + j12 - jj16$$

$$= 9 + 16$$

$$= 25$$

$$\text{Magnitude} = \sqrt{25} = 5.$$

You will recall that 5 is a real number, equal to the number of toes on your left foot. You don't have to trust me; just count them. We will return to this important point in IWU #35.

The *phase angle* is an angle whose tangent is the ratio of the imaginary to the real, which in this example is an angle whose tangent is 4/3. Instead of writing the phrase "an angle whose tangent is" people have agreed to abbreviate this as arctangent, arctan, ATAN, or $\tan^{-1}$. So, here we would say that the phase angle is arctan(4/3). Your calculator will tell you that this is about 36.87 degrees.

In general, we can write a complex number as,

$$z = x + jy$$

and its complex conjugate is usually written as $z*$,

$$z^* = x - jy$$

so the square of its magnitude is,

$$|z|^2 = zz^* = x^2 + y^2$$

so the magnitude is,

$$|z| = \sqrt{x^2 + y^2}$$

and its phase angle is,

$$\theta = \arctan\left(\frac{y}{x}\right).$$

When a complex number is written as $x + jy$ we say it is in Cartesian or rectangular form. When it is written as a magnitude and phase angle we say it is in polar form. Cartesian coordinates are named for René Descartes (1596-1650), French philosopher, soldier, scientist, and mathematician. Among his many accomplishments he invented analytic geometry which helped us all in trying to learn calculus.

Once upon a time there lived in France a fairly intelligent horse. The horse was particularly good at mathematics. Algebra, vector analysis, and calculus were rather easy for the horse to learn, but it could never seem to understand analytic geometry. The moral of the story is: Don't put Descartes before de horse.

Now you can do addition, subtraction, multiplication, division, powers and roots, and logs. If you accept all of this you are ready for a most amazing thing. (See IWU #35.)

IWU #35. The Magic of the Mysterious Numbers

Euler was a Swiss mathematician who lived from 1707-1783. His productivity was promethean and is even more remarkable because much of his work was done after he became blind in one eye in 1735 and totally blind in 1766. His insight was greater than his outsight. Euler seemed to have the Midas touch in calculus, geometry, algebra, number theory, and probability. He contributed to applied mathematics in the fields of acoustics, optics, mechanics, astronomy, artillery, navigation, statistics, and finance; a good man to have on your team. In the twentieth century a Swiss scientific society started publishing his complete works. Fifty years later, forty-seven volumes had been published and the project was still not finished.

Abraham de Moivre, a French mathematician, apparently conceived an idea upon which Euler elaborated. The result is one of the most useful, practical expressions applied daily by physicists, engineers, and ordinary mortals without hesitation, almost intuitively, and yet it is without a doubt one of the most mystical expressions we have in our possession.

Those who have made careers of studying famous formulas say that it is the most famous of all. The relationship is, in general,

$$e^{j\theta} = \cos\theta + j\sin\theta$$

Here we have a startling relationship between the imaginary and the transcendental. But there is more: For the special case when $\theta = \pi$,

$$e^{j\pi} = -1$$

or

$$e^{j\pi} + 1 = 0$$

There is nothing else like it in the annals of mathematical mythology. Any way you look at it, it is amazing. Here is an equation involving quantities, seemingly so different, that were so crucial in the development of mathematics; and they are all related! So what was the fight all about? You accept one, you accept all. Two transcendental numbers and an imaginary unit make a negative integer! Two transcendental numbers, an imaginary unit, and a positive integer combine to give zero! Nothing! That is what we call an IWU.

We're talking big-time magic here. You can make all of the real positive integers from a negative real transcendental and an imaginary transcendental. The real positive integers are the ones we know and love so well, the ones we were so very sure of and thought we could trust when all other relationships turn to dust, the ones we thought we could count on, the ones we *can* count on our fingers and toes. It's as easy as 1, 2, 3.

$$-e^{j\pi} = 1$$

$$-e^{j\pi} - e^{j\pi} = 2$$

$$-e^{j\pi} - e^{j\pi} - e^{j\pi} = 3$$

and on to infinity. You can't trust anything any more. We'll pause while you reset your circuit breakers. According to legend, Professor Benjamin Peirce, one of Harvard's famous mathematicians of the nineteenth century, told his students one day, "Gentlemen, that is surely true, it is absolutely paradoxical; we cannot understand it, and we don't know what it means, but we have proved it, and therefore, we know it must be the truth."

There once was a student at Trinity
Who found the square root of infinity
But it gave him such fidgets
To count up the digits
He quit Math and took up Divinity.

So, now we can write a complex number z as

$$z = r\,e^{j\theta}$$

where

$$r = |z|$$

and

$$\theta = \arctan\left(\frac{\mathrm{Im}\ z}{\mathrm{Re}\ z}\right)$$

because

$$z = r\,e^{j\theta} = r\,(\cos\theta + j\sin\theta) = x + jy$$

So, Pilgrim, you say you just want to have fun. OK, let's take the natural logarithm of z. It's easier in polar form. Ready?

$$\ln(z) = \ln(r\,e^{j\theta}) = \ln(r) + j\theta\ .$$

Ladies and gentlemen, you have just seen The Wizard of θz take the natural logarithm of a complex number right before your very eyes!

What about $z*$?

$$\ln(z^{*}) = \ln(r\,e^{-j\theta}) = \ln(r) - j\theta\ .$$

So what is the natural log of $zz*$?

$$\ln(zz^{*}) = \ln(z) + \ln(z^{*})$$
$$= \ln(r) + j\theta + \ln(r) - j\theta$$
$$= 2\ln(r) = \ln(r^{2})$$

so,

$$\ln(zz^{*}) = \ln(r^{2})$$

so,

$$zz^{*} = r^{2}.$$

But we already knew that. Nevertheless, it is comforting to hear once more the old sayings your grandmother taught you.

$$zz^{*} = |z|^{2} = r^{2}.$$

Now we can use this to obtain a few particularly interesting expressions for $\ln(z)$, because we know that $\ln(z) = \ln(r) + j\theta$.

$$\ln(z) = \ln(|z|) + j\theta$$

and

$$\ln(z) = \ln[(zz^*)^{\frac{1}{2}}] + j\theta$$

so

$$\ln(z) = 0.5 \ln(zz^*) + j\theta$$

Now we can unmix the polar and Cartesian forms by writing,

$$z = Re(z) + jIm(z)$$

where

$$x = Re(z) \quad \text{and} \quad y = Im(z)$$

so,

$$\ln(z) = 0.5 \ln(zz^*) + j \arctan\left(\frac{Im\ z}{Re\ z}\right)$$

"Toto, I've a feeling we're not in Kansas any more."

from the 1939 MGM film *The Wizard of Oz*
based on the book by L. Frank Baum

As we approach the Emerald City in the Wonderful Land of θz things become even stranger. Because $j = e^{j\pi/2}$ we can take the natural log of both sides and obtain,

$$\ln(j) = j\frac{\pi}{2}$$

so,

$$\frac{1}{j}\ln(j) = \frac{\pi}{2}$$

and so,

$$\ln(j^{1/j}) = \frac{\pi}{2}$$

We just showed that $\pi/2$ is the natural log of the jth root of j. Whatever that means.

Complex Operations in a Spreadsheet

Some spreadsheet versions have a convenient set of "@-functions" that perform complex operations (in *Excel* the "@" is not used). *Excel* (4.0 and later) and *Quattro Pro* (5.0 and later) contain the FFT, IFFT, and the following :

@COMPLEX: Converts real and imaginary parts into a complex number.
@IMABS: Returns the magnitude (or norm) of a complex number.
@IMAGINARY: Returns the imaginary part of a complex number.
@IMARGUMENT: Returns the argument θ, an angle in radians.
@IMCONJUGATE: Returns the complex conjugate of a complex number.
@IMCOS: Returns the cosine of a complex number.
@IMDIV: Returns the quotient of two complex numbers.
@IMEXP: Returns the exponential of a complex number.
@IMLN: Returns the natural logarithm of a complex number.
@IMLOG10: Returns the base-10 logarithm of a complex number.
@IMLOG2: Returns the base-2 logarithm of a complex number.
@IMPOWER: Returns a complex number raised to a complex power.
@IMPRODUCT: Returns the product of two complex numbers.
@IMREAL: Returns the real part of a complex number.
@IMSIN: Returns the sine of a complex number.
@IMSQRT: Returns the square root of a complex number.
@IMSUB: Returns the difference of two complex numbers.
@IMSUM: Returns the sum of complex numbers.

Impedance and Admittance

These are extraordinarily useful quantities in the description of physical systems of all sorts simply because they express relations between cause and effect. It is appropriate that we discuss them here because in general they are complex quantities. Let us start with impedance, which we have seen before, and which is just the ratio of cause to effect. To be specific let the cause be V and the effect be I. It should be clear that this discussion is completely general and is not restricted to electrical circuits. The impedance is,

$$Z = \frac{V}{I} = \frac{Cause}{Effect} = \frac{Stimulus}{Response}$$

so for electrical circuits,

$$V = IZ.$$

Often we are interested in writing an expression for the effect,

$$I = \frac{V}{Z} \ .$$

Because Z is complex it would be convenient to multiply V by the reciprocal of Z. The reciprocal of Z is called the admittance and is represented by Y. Don't ask me why.

$$Y = \frac{1}{Z} \ .$$

Because Z is complex we can write it in cartesian form in terms of its real and imaginary parts,

$$Z = R + jX \ .$$

As usual R is the resistance and X is the net reactance. The reciprocal of Z is,

$$\frac{1}{Z} = \frac{1}{R+jX} = \frac{R-jX}{(R+jX)(R-jX)} = \frac{R}{R^2+X^2} - j\frac{X}{R^2+X^2}$$

So, now we have Y written in cartesian form,

$$Y = \frac{R}{R^2+X^2} - j\frac{X}{R^2+X^2} \ .$$

The real part is called the *conductance* G and the imaginary part is called the *susceptance B*.

$$G = \frac{R}{R^2+X^2}$$

$$B = -j\frac{X}{R^2+X^2}$$

As usual we will emphasize the obvious:

- In Z the resistance and reactance are completely separated into the real and imaginary parts.
- In Y the resistance and reactance become mixed, so that *both R and X* appear in the real and imaginary parts.
- The conductance is simply $1/R$ when $R^2 >> X^2$.
- The susceptance is simply $1/X$ when $X^2 >> R^2$.
- The imaginary part of Y has its sign opposite to that of the imaginary part of Z. As a result, if the phase angle of Z is positive then the phase angle of Y will be negative.

Digging Deeper

A History of π, Fifth Ed, P. Beckmann (Golem Press, Boulder, Colorado, 1982). ISBN 0-911-76218-3. This is a short history of a number with a long history.

Mathematics and the Imagination, E. Kasner and J. Newman (Simon and Schuster, New York, 1943). A classic. Among the gems, Chapter 3 discusses π, *e*, and *j*.

The History of the Mysterious Numbers, P. Dubreil, in *Great Currents of Mathematical Thought*, Vol. I, F. LeLionnais, Editor (Dover Publications, New York, 1971). ISBN 0-468-62723-3. Mathematics from the French perspective.

The Story of a Number, E. Maor (Princeton University Press, Princeton, New Jersey, 1994). ISBN 0-691-03390-0. This book is devoted to one of our favorite transcendental numbers, *e*.

Fundamentals of Complex Analysis for Mathematics, Science, and Engineering, 2nd Edition, E. B. Saff and A. D. Snider (Prentice Hall, Englewood Cliffs, New Jersey, 1993). ISBN 0-13-327461-6. This textbook covers the field at the undergraduate level and it has many examples of applications.

f(z), *The Complex Variables Program* (Lascaux Graphics, 3771 E. Guthrie Mtn. Place, Tucson, Arizona 85718) Phone (800) 338-0993. This program lets you visualize things that previously you could only dream about. It is available for the PC and the Macintosh.

Appendix 5

Details of Simple Filters

Being myself a remarkably stupid fellow, I have had to unteach myself the difficulties, and now beg to present...the parts that are not hard. Master these thoroughly, and the rest will follow.

Professor Silvanus P. Thompson, F.R.S.

What This Appendix is About

In this Appendix we will show you how simple filters work. The world is full of complicated analog and digital filters that accomplish remarkable feats of signal processing. But you have to start somewhere and if you take the time to master the easy parts thoroughly the rest will follow.

We are going to discuss four basic analog filters from the viewpoint of a voltage divider that depends on frequency. To get in the mood let's review the idea of a voltage divider using only two resistors.

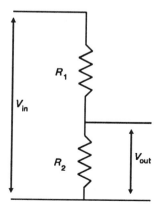

Fig. A5-1. This is not a filter. It is a voltage divider using two resistors, so the output does not depend on frequency.

The filters that we will discuss here are "passive" rather than "active." This means that no energy is added to the signal by the filter; there is no amplification and no feedback. Because of energy loss in the filters the outputs will always be less than the inputs. Because energy is conserved the input voltage V_{in} must be equal to the sum of the IR voltage drops $V_{in} = IR_1 + IR_2$.

The output voltage V_{out} is the voltage drop across IR_2. We are interested in the Gain, the ratio of V_{out}/V_{in}. Divide the equation above by IR_2 and you get,

$$\frac{V_{in}}{IR_2} = \frac{IR_1}{IR_2} + 1 \ .$$

The current I cancels in the first term on the right so this can be re-written as,

$$\frac{V_{in}}{V_{out}} = \frac{R_1}{R_2} + 1 = \frac{R_1 + R_2}{R_2} \ .$$

Finally, take the reciprocal of this to find the Gain,

$$G = \frac{R_2}{R_1 + R_2} \ .$$

The Gain of the resistive voltage divider is clearly independent of frequency, and depends only on the values of the resistances. In fact the Gain only depends on the *ratio* of the resistances as you can see by doing a little algebra on the equation above,

$$G = \frac{R_2/R_1}{1 + R_2/R_1} \ .$$

You can regard this as the output resistance normalized to R_1. The Gain as a function of the normalized output resistance is shown in Fig. A5-2.

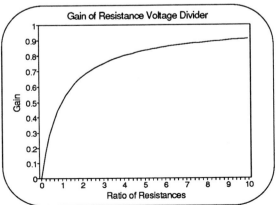

Fig. A5-2. Gain of the resistive voltage divider as a function of R_2/R_1. Gain is independent of frequency.

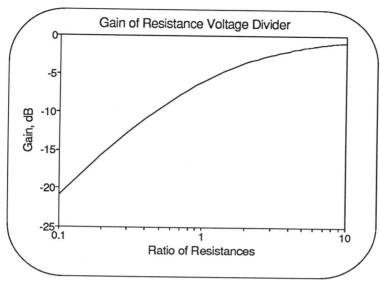

Fig. A5-3. Gain of the resistance voltage divider in dB, with R_2/R_1 on a logarithmic scale. These are the same data shown in Fig. A5-2.

To prepare for what's to come, it will also useful to view the Gain of the resistance voltage divider in a log-log representation in terms of dB and a logarithmic resistance scale, shown in Fig. A5-3.

Resistance, Capacitance, and Inductance

There are at most three types of passive elements in these filters: R, C, and L. In what follows the resistance R is expressed in ohms, capacitance C in farads, and inductance L in henrys. These are fundamentally defined in terms of how they relate voltage, charge, current, and time rate of change of current, shown in Table A5-1.

Table A5-1. Definitions of Passive Elements		
Element	Unit	Defining Equation
C	farad	$C = q/V$
R	ohm	$R = V/I$
L	henry	$V = -L\Delta I/\Delta t$

In the Table q is electric charge in coulombs, V is potential difference in volts, I is current in amperes, and $\Delta I/\Delta t$ is the time rate of change of current. Resistance transforms electrical energy into thermal energy, capacitance stores electrical energy in the electric field, and inductance stores electrical energy in the magnetic field. When these components are combined in various ways their combined effects may appear to be strange and wonderful, but basically all they are doing is storing some energy and wasting some energy.

Sinusoidal Response

Now that you have these basics let's confine our attention to voltages and currents that vary in time only in a sinusoidal way. In other words, we are going to discuss the systems in terms of their response to a pure tone. This introduces a great simplification and if we can find the response for a sine then by Fourier analysis and synthesis we can find the response to any kind of input.

We are going to use the methods of IWU #17 and IWU #30 in Appendix 4. It is not mandatory that we do it this way. We could do everything with sines and cosines but we will use the methods of the Wizard of θZ because they are simpler and easier.

In linear systems the fundamental relation between voltage V, current I, and impedance Z is given by Ohm's law,

$$V = IZ.$$

In Ohm's law for a.c. circuits Z is the complex impedance (measured in Ohms), which is the ratio of cause (V) to effect (I),

$$Z = R + jX.$$

Here X called the *reactance*, which is the imaginary part of the impedance. It is the sum of the imaginary parts X_L and X_C. Resistance has no imaginary part, so its impedance is entirely real. Ideal capacitance and inductance are purely imaginary (pure reactance) with no real part. Reactance X has units of Ohms. A sinusoidal voltage across an ideal inductor leads the current by 90°. The voltage across an ideal capacitor lags the current by 90°. All of this is summarized in Table A5-2.

Table A5-2. Reactance and Resistance	
Element	*Reactance, Resistance (Ohms)*
C	$X_C = 1/j\omega C = -j/\omega C$
R	R
L	$X_L = j\omega L$

These elements can be represented as vectors in the complex plane, as we did in Appendix 4. This is shown graphically in Fig. A5-4, in which the magnitude of the impedance and the phase angle are also shown.

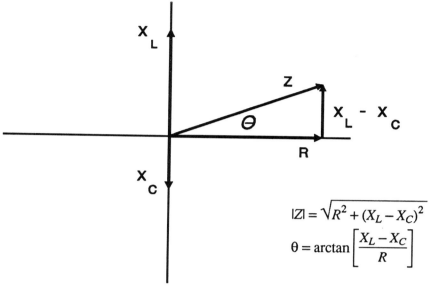

$$|Z| = \sqrt{R^2 + (X_L - X_C)^2}$$

$$\theta = \arctan\left[\frac{X_L - X_C}{R}\right]$$

Fig. A5-4. Inductive reactance, resistance, and capacitative reactance are represented as vectors in the complex plane. The total impedance Z is the sum of these vectors. The sum depends on frequency in accordance with Table A5-2.

You also need to know that energy is conserved so that the voltage at the input is equal to the sum of all of the voltages around a closed path. That's about all you need to know for now.

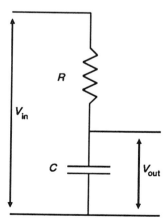

Fig. A5-5. A frequency-dependent voltage divider connected as a low-pass filter. The low frequencies dominate the output because the voltage across the capacitor decreases as frequency increases.

Low-Pass Filter

Suppose we connect a resistor and a capacitor as shown in Fig. A5-5. This will make a voltage divider that depends on frequency. Let's add up all the voltages:

$$V_{in} = V_r + V_c.$$

By Ohm's law we can write this as,

$$V_{in} = IR + IX_c = IR + I\frac{1}{j\omega C}.$$

Now this is what we are going to do:

- Calculate the ratio of the output voltage to the input voltage, which is the voltage gain. This will be a complex function because of the capacitor.

- Calculate the phase angle θ by means of the arctangent of the ratio of the imaginary part to the real part.

- Calculate the magnitude of the voltage gain from the square root of the square of the complex gain.

The output voltage V_{out} is just the voltage across the capacitor so we can divide the equation above by $I/j\omega C$ and then re-write the equation as,

$$\frac{V_{in}}{V_{out}} = \frac{IR}{I/j\omega C} + 1 = j\omega RC + 1.$$

We are interested in the reciprocal of this, because the reciprocal is the complex gain. We are going to represent the gain as $G(j\omega)$ to emphasize its complex nature:

$$G(j\omega) = \frac{V_{out}}{V_{in}} = \frac{1}{1 + j\omega RC}.$$

This might be enough information for a mathematician but ordinary mortals usually need to put it in a form that is more easily recognized. Let's multiply the numerator and denominator by the complex conjugate as we did in Appendix 4:

$$G(j\omega) = \frac{1 - j\omega RC}{1 + (\omega RC)^2}.$$

The real part of the complex gain is $1/[1 + (\omega RC)^2]$ and the imaginary part is $-j\omega RC /[1 + (\omega RC)^2]$. Now we can find the phase angle,

$$\theta(\omega) = \arctan\left(\frac{Im}{Re}\right) = \arctan(-\omega RC).$$

The magnitude of the Gain is simply the square root of $|G(j\omega)|^2$ and the easiest way to find this is as follows:

$$|G(j\omega)|^2 = \frac{1}{(1+j\omega RC)(1-j\omega RC)} = \frac{1}{1+(\omega RC)^2} \, .$$

We will have the magnitude of the gain if we take the square root of this expression.

$$|G(j\omega)| = \frac{1}{\sqrt{1+(\omega RC)^2}} \, .$$

We need to take the log of this equation and multiply it by 20 to get it in dB:

$$dB = 20 \log \left(\frac{1}{\sqrt{1+(\omega RC)^2}} \right) = -10 \log \left(\sqrt{1+(\omega RC)^2} \right) \, .$$

That's all there is to it. Now we have the Gain (in dB) and the phase of a simple low-pass filter. See Fig. A5-6 and Fig. 6-1 for Bode plots of the Gain and phase.

Let's summarize our results:

- At low frequencies the Gain and the phase shift are almost 0 dB and 0°.

- As the frequency increases the Gain decreases without limit and the phase shift asymptotically approaches $-90°$.

- At the corner frequency ($\omega_c = 1/RC$) the Gain is -3 dB and the phase is $-45°$.

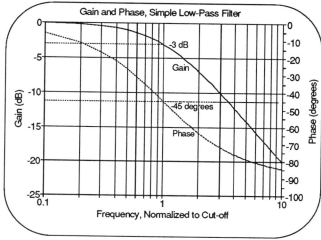

Fig. A5-6. Bode plot of a single-pole low-pass filter, with cutoff frequency normalized to 1. Note −3dB gain and −45° phase shift at cutoff frequency.

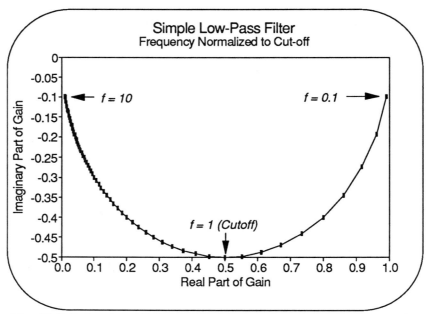

Fig. A5-7. Polar plot of Gain with frequency as a parameter. The cutoff frequency is normalized to 1. A vector drawn from the upper left-hand corner to the curve represents the Gain at any frequency. Phase goes negative as frequency increases. $\Delta f = 0.1$ between markers.

Another useful representation of the frequency response of a filter can be obtained by plotting the imaginary part of the complex Gain as a function of the real part, shown in Fig. A-7. This polar plot is similar to the Argand diagram used in complex variables. In this representation you can see the "shrink and turn" effect of the filter. For a low-pass filter the magnitude shrinks and the phase turns more negative as the frequency increases.

High-Pass Filter

Next, let's exchange the positions of R and C in Fig. A5-4 to make a high-pass filter as shown in Fig. A5-8. Now the output is across the resistor instead of the capacitor. Let's add up all the voltages in Fig. A5-8:

$$V_{in} = V_c + V_r .$$

By Ohm's law we can write this as,

$$V_{in} = IX_c + IR = I\frac{1}{j\omega C} + IR .$$

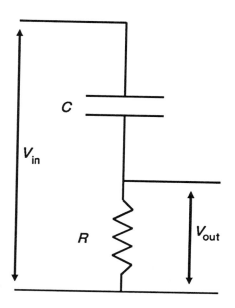

Fig. A5-8. A simple high-pass filter made with a capacitor and a resistor. The output voltage increases with frequency because the capacitor passes the high frequencies and blocks the low frequencies.

Now this is what we are going to do:

- Calculate the ratio of the output voltage to the input voltage, which is the voltage gain. This will be a complex function because of the capacitor.

- Calculate the phase angle θ by means of the arctangent of the ratio of the imaginary part to the real part.

- Calculate the magnitude of the voltage gain from the square root of the square of the complex gain.

The output voltage V_{out} is just the voltage across the resistor so we can divide the equation above by IR and then re-write the equation as,

$$\frac{V_{in}}{V_{out}} = \frac{IR}{IR} + \frac{I}{IR(j\omega C)} = 1 + \frac{1}{j\omega RC}.$$

We are interested in the reciprocal of this, because the reciprocal is the complex gain. We are going to represent the gain as $G(j\omega)$ to emphase its complex nature:

$$G(j\omega) = \frac{V_{out}}{V_{in}} = \frac{j\omega RC}{1 + j\omega RC}.$$

As with the low-pass filter this might be enough information for a mathematician but we need to put it in a form that is more easily recognized. To separate the real and imaginary parts let's multiply the numerator and denominator by $1 - j\omega RC$ as before:

$$G(j\omega) = \frac{j\omega RC\,(1 - j\omega RC)}{1 + (\omega RC)^2}\,.$$

So, as before we find the real part is $(\omega RC)^2/[1 + (\omega RC)^2]$ and the imaginary part is $j\omega RC/[1 + (\omega RC)^2]$. Now we can find the phase angle,

$$\theta(\omega) = \arctan\!\left(\frac{\text{Im}}{\text{Re}}\right) = \arctan\!\left(\frac{1}{\omega RC}\right)\,.$$

The magnitude of the Gain is simply the square root of $|G(j\omega)|^2$ and the easiest way to find this is as follows:

$$|G(j\omega)|^2 = \frac{(j\omega RC)(-j\omega RC)}{(1 + j\omega RC)(1 - j\omega RC)} = \frac{(\omega RC)^2}{1 + (\omega RC)^2}\,.$$

We will have the magnitude of the gain if we take the square root of this expression.

$$|G(j\omega)| = \frac{\omega RC}{\sqrt{1 + (\omega RC)^2}}\,.$$

We need to take the log of this equation and multiply it by 20 to get it in dB:

$$\text{dB} = 20\log\!\left(\frac{\omega RC}{\sqrt{1 + (\omega RC)^2}}\right) = 10\log\!\left(\frac{(\omega RC)^2}{1 + (\omega RC)^2}\right)\,.$$

We now have the Gain (in dB) and phase of a simple high-pass filter (see Fig. A5-7). Let's summarized what we have found:

- At high frequencies the Gain and the phase shift are almost 0 dB and 0°.

- As the frequency decreases the Gain decreases without limit and the phase shift asymptotically approaches +90°.

- At the corner frequency ($\omega_c = 1/RC$) the Gain is –3 dB and the phase is +45°.

The magnitude and phase response is shown in Fig. A5-9. Compare this with Fig. A5-6 for the low-pass filter. Let's finish the high-pass filter by examining its polar diagram (Fig. A5-10) which shows the characteristic "stretch and turn" of the vector Gain as frequency increases. It might be instructive for you to draw a vector from the origin to the curve at the cutoff frequency. Observe that the magnitude of this vector is $\sqrt{0.5^2 + 0.5^2} = 0.707$ and the phase is +45°.

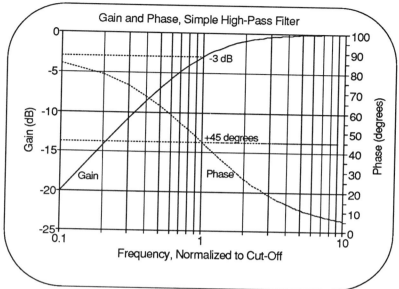

Fig. A5-9. Bode plot of a single-pole high-pass filter, with frequency normalized to the cut-off frequency.

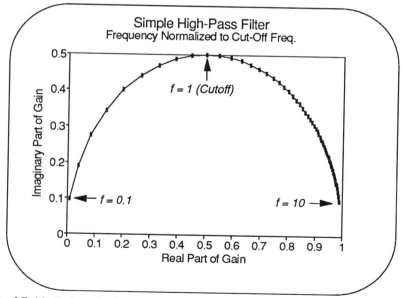

Fig. A5-10. Polar plot for a single-pole high-pass filter. At the cutoff frequency (normalized to $f_c = 1$) the real part is equal to the imaginary part, and the phase angle is +45°. $\Delta f = 0.1$ between markers.

Band-Pass Filter

There are two general ways that we can construct a band-pass filter:

- Make a tuned circuit that will *block* all but a narrow band of frequencies. The filter will provide a high series impedance to all frequencies except those in the pass band.

- Make a tuned circuit that will *bypass* (to ground) all but a narrow band of frequencies. The filter will provide a low series impedance (to ground) to all frequencies except those in the pass band.

We are going to choose the second procedure because it has practical advantages of placing one end of the capacitor and inductor at ground potential. This is a good safety precaution when high voltages are involved, as in radio transmitters. This procedure also minimizes the effect of hand capacitance when you touch a tuning knob in a variable filter.

The capacitor and inductor are connected in parallel so their impedance will be high on resonance and lower at other frequencies. The Q of the circuit will determine the width of the pass band.

$$V_{in} = IR + IZ.$$

Before we can proceed we have to complete the auxiliary task of finding Z for the capacitor and inductor in parallel. Here, Z is the impedance of the parallel capacitor and inductor which can be found using a little algebra like this.

$$\frac{1}{Z} = \frac{1}{Z_C} + \frac{1}{Z_L}$$

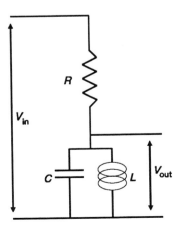

Fig A5-11. A band-pass filter using a capacitor and inductor in parallel.

Let's add up all the voltages in Fig. A5-11:

$$\frac{1}{Z} = \frac{1}{1/j\omega C} + \frac{1}{j\omega L} = j\omega C + \frac{1}{j\omega L}$$

$$\frac{1}{Z} = \frac{(j\omega C)\,(j\omega L) + 1}{j\omega L} = \frac{1 - \omega^2 LC}{j\omega L}\; .$$

Finally, Z is the reciprocal of this expression,

$$Z = \frac{j\omega L}{1 - \omega^2 LC}\; .$$

So we can write,

$$V_{in} = IR + I\frac{j\omega L}{1 - \omega^2 LC}$$

The second term is the output voltage V_{out} so we divide the entire equation by this,

$$\frac{V_{in}}{V_{out}} = \frac{IR}{I\dfrac{j\omega L}{1 - \omega^2 LC}} + 1\; .$$

As before (and always) the current I cancels and we have,

$$\frac{V_{in}}{V_{out}} = \frac{R\,(1 - \omega^2 LC)}{j\omega L} + 1\; .$$

This can be slightly re-arranged for later convenience,

$$\frac{V_{in}}{V_{out}} = \frac{(1 - \omega^2 LC) + j\omega L/R}{j\omega L/R}\; .$$

The reciprocal of this is the Gain, $G(j\omega)$,

$$G(j\omega) = \frac{V_{out}}{V_{in}} = \frac{j\omega L/R}{(1 - \omega^2 LC) + j\omega L/R}\; .$$

This is difficult to interpret until we clear the denominator of j's so we can separate the Gain into its real and imaginary parts. Multiply the numerator and denominator by the complex conjugate of the denominator,

$$G(j\omega) = \frac{V_{out}}{V_{in}} = \frac{(j\omega L/R)\,[(1 - \omega^2 LC) - j\omega L/R]}{[(1 - \omega^2 LC) + j\omega L/R]\,[(1 - \omega^2 LC) - j\omega L/R]}\; .$$

$$G(j\omega) = \frac{(\omega L/R)^2 + (j\omega L/R)(1 - \omega^2 LC)}{(1 - \omega^2 LC)^2 + (\omega L/R)^2} .$$

Now that we have the real part separated from the imaginary part we can immediately find the phase angle (as before),

$$\theta(\omega) = \arctan\left(\frac{\text{Im}}{\text{Re}}\right) = \arctan\left(\frac{1 - \omega^2 LC}{\omega L/R}\right) .$$

Finally we need to calculate the magnitude of the Gain.

$$|G(j\omega)|^2 = G^*(j\omega)\, G(j\omega) = \frac{(-j\omega L/R)(+j\omega L/R)}{[(1-\omega^2 LC) - j\omega L/R][(1-\omega^2 LC) + j\omega L/R]}$$

or,

$$|G(j\omega)|^2 = \frac{(\omega L/R)^2}{(1 - \omega^2 LC)^2 + (\omega L/R)^2} .$$

The magnitude of the Gain is the square root of the equation above,

$$|G(j\omega)| = \frac{(\omega L/R)}{\sqrt{(1 - \omega^2 LC)^2 + (\omega L/R)^2}} .$$

Convert this to dB as usual by taking the log and multiplying by 20,

$$dB = 20 \log\left(\frac{(\omega L/R)}{\sqrt{(1 - \omega^2 LC)^2 + (\omega L/R)^2}}\right) .$$

So, we have the Gain and phase of a band-pass filter. That wasn't so bad, was it? The Bode plot of the Gain and phase is shown in Fig. 7-1, Chapter 7. Remember the quantity $\omega L/R$? At the resonance frequency that's what we called Q in Chapters 4 and 5. However, here this is not Q because in this filter R is not the resistance in the filter! Here, in fact, we have assumed that the resistance of L is zero. While you're resting carry out the calculation above but include a small resistance r in the inductor. You can follow the method shown below for the band-stop filter in which a finite resistance is used.

Band-Stop Filter

There are two general ways that we can construct a band-stop filter:

- Make a tuned circuit that will *pass all but a narrow band* of frequencies to the output. The filter will provide a low series impedance to the output for all frequencies except those in the stop band.

- Make a tuned circuit that will *bypass (short-circuit to ground) a narrow band* of frequencies. The filter will provide a high series impedance (to ground) for all frequencies except those in the stop band.

Again we are going to choose the second procedure because it has the practical advantages of safety and minimizing hand capacitance. The capacitor is grounded so its capacitance doesn't change when you touch it for tuning.

The circuit that we are going to analyze looks like that shown below. The capacitor and inductor are connected in series so their impedance will be low on resonance and higher at other frequencies. We are going to be realistic and include a small internal resistance r in the inductor. The Q of the circuit will determine the width of the stop band and the phase characteristic. Let's add up all the voltages in Fig. A5-12:

$$V_{in} = IR + IZ.$$

Here Z is the impedance of the capacitor and inductor in series, which is simply the sum of the impedances,

$$Z = Z_C + Z_L = \frac{1}{j\omega C} + (j\omega L + r).$$

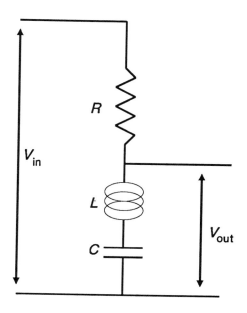

Fig. A5-12. A band-stop filter using a capacitor and inductor in series. The small internal resistance r in L is not shown.

With a little algebra we can simplify this and get,

$$Z = \frac{1 - \omega^2 LC + j\omega rC}{j\omega C}.$$

The output voltage V_{out} is IZ and the input voltage V_{in} is $IR + IZ$ so let's divide V_{out} by V_{in} to get the Gain,

$$\frac{V_{out}}{V_{in}} = \frac{IZ}{IR + IZ} = \frac{\dfrac{1 - \omega^2 LC}{j\omega C}}{R + \dfrac{1 - \omega^2 LC + j\omega rC}{j\omega C}}$$

After a little simplification the complex Gain can be written as,

$$G(j\omega) = \frac{V_{out}}{V_{in}} = \frac{(1 - \omega^2 LC) + j\omega rC}{(1 - \omega^2 LC) + j\omega(R + r)C}.$$

Once again we need to clear the denominator of j's in order to separate $G(j\omega)$ into its real and imaginary parts. Multiply the numerator and denominator by the complex conjugate and you will obtain,

$$G(j\omega) = \frac{(1 - \omega^2 LC) + j\omega rC}{(1 - \omega^2 LC) + j\omega(R + r)C} \frac{(1 - \omega^2 LC) - j\omega(R + r)C}{(1 - \omega^2 LC) - j\omega(R + r)C}.$$

$$G(j\omega) = \frac{(1 - \omega^2 LC)^2 + \omega^2 r(R + r)C^2 - j\omega RC(1 - \omega^2 LC)}{(1 - \omega^2 LC)^2 + [\omega(R + r)C]^2}.$$

We can now find the phase angle by the ratio of the imaginary to the real,

$$\theta(\omega) = \arctan\left(\frac{Im}{Re}\right) = \arctan\left(\frac{-\omega RC(1 - \omega^2 LC)}{(1 - \omega^2 LC)^2 + \omega^2 r (R + r)C^2}\right).$$

Finally we need to find the magnitude of the Gain, using our standard operating procedure.

$$G(j\omega)G^*(j\omega) = \frac{[(1 - \omega^2 LC) - j\omega rC] \, [(1 - \omega^2 LC) + j\omega rC]}{[(1 - \omega^2 LC) + j\omega(R + r)C] \, [(1 - \omega^2 LC) - j\omega(R + r)C]}.$$

After carrying out the multiplication we find,

$$|G(j\omega)|^2 = \frac{(1 - \omega^2 LC)^2 + (\omega rC)^2}{(1 - \omega^2 LC)^2 + [\omega(R + r)C]^2}.$$

The magnitude of the Gain is the square root of the equation above,

$$|G(j\omega)| = \left[\frac{(1 - \omega^2 LC)^2 + (\omega rC)^2}{(1 - \omega^2 LC)^2 + [\omega(R + r)C]^2}\right]^{1/2}.$$

Take the log of this and multiply by 20 to get the Gain in dB,

$$dB = 10 \log_{10}\left[\frac{(1 - \omega^2 LC)^2 + (\omega rC)^2}{(1 - \omega^2 LC)^2 + [\omega(R + r)C]^2}\right].$$

The factor 10 is due to fact that $20 \log x^{1/2} = 10 \log x$. The Bode plot of Gain and phase for this filter is shown in Fig. 7-7, Chapter 7.

Connections between the Time and Frequency Domains

In this Appendix we have concentrated on the Gain and phase properties in the frequency domain. In Chapters 6 and 7 we also showed the time domain properties (the waveforms) in various representations of amplitude vs. time and output vs. input. Obviously there must be a connection between these domains; if you know the characteristics in one domain you must also know them in the other. The connection is the Fourier transform (Chapter 8). The FFT and IFFT operations provide the bridge between time and frequency, or any two variables related by the transform. A fuller explanation involves a little calculus. In the Preface we promised not to use any, so we must stop.

Digging Deeper

HF Filter Design & Computer Simulation, R. W. Rhea (Noble Publishing, Tucker, Georgia, 1994). ISBN 1-884932-25-8. This is a practical book that guides you from concept through hardware realization and measurement of filter properties. The emphasis is on high-frequency filters but the general ideas are universal.

Digital Signal Processing Experiments, A. Kamas and E. A. Lee (Prentice Hall, Englewood Cliffs, New Jersey, 1989). ISBN 0-13-212853-5. This book and its included software turns your computer into an easy-to-use digital signal processing lab. The software is a subset of the *DSPlay* program produced by the Burr-Brown Corportation.

Appendix 6

Bits, Digits, and dB

The resolution of a digital voltmeter is usually expressed in terms of the number of digits that you can measure with confidence. On the other hand, the resolution of an analog/digital converter or a digital/analog converter is specified in terms of the number of bits (binary digits). These two measures of resolution are related, and both can be expressed in terms of the dynamic range of the instrument, in dB. They are shown in Table A6-1, for a fixed-point number system, with no allowance for a sign bit.

Table A6-1. Bits, Digits, and dB		
Bits	Digits	dB
8	2.5	48.2
12	3.5	72.2
16	4.5	96.3
18	5.5	108.4
20	6.0	120.4
22	6.5	132.5
24	7.0	144.5
26	7.5	156.5
28	8.5	168.6
30	9.5	180.6
32	10.0	192.7

The bottom entry is near the limit of current A/D technology. When signals are digitized the resolution, dynamic range, and S/N depend on the number of bits available in the analog-to-digital conversion, and the type of number system.

Digital Resolution

The digital resolution depends on the number of bits in which a digitzed value is stored. This represents the number of divisions into which the full-scale range is divided, For example, a 0-10 V range with a 12-bit resolution will have 4096 (or 2^{12}) divisions of 2.44 mV each (that is, 10/4096). A resolution of 16 bits corresponds to 0.15 mV for each division (or $10/2^{16} = 10/65,536$).

Absolute Resolution

Absolute resolution refers to the case in which it is necessary to maintain a definite accuracy when representing a data point. For example, suppose your data range from 0 to 1.0 V. Absolute resolution is equivalent to the requirement that you be able to resolve with confidence 1 mV or perhaps 10 μV, throughout the entire range. If the data range from 0 to 10 V the same accuracy of 1 mV or 10 μV is still retained.

Relative Resolution

Relative resolution describes the case in which it is necessary to keep errors below a certain level (conveniently measured in dB) relative to a given number. For example, in maintaining a constant signal-to-noise (S/N) ratio the level of acceptable noise can increase as the signal increases. For data ranging from 0 to 1.0 V a 1% resolution is equivalent to resolving 0.01 V or 10 mV, and a 0.01% resolution requires resolving 100 μV. If the data range from 0 to 10 V then the corresponding resolutions are 100 mV and 1 mV.

Fixed Point and Floating Point Numbers

The *dynamic range* can be conveniently expressed in dB, in terms of the ratio of the largest to the smallest value of data that can be accurately represented. In the early days of digital signal processing people used a fixed-point number system, for technical reasons. This was typically implemented in 16 bits which limited the application of processing in many situations. Now the added complexity of floating-point numbers is not a problem, and this is typically implemented in 32 bits.

Fixed-Point Numbers

For example, a 16-bit fixed-point number system with the sign (+ or −) as the most significant bit has $2^{15} - 1$ as its largest number and 1 as its smallest number. However, since the number's least significant bit changes when an analog input reaches half the value of the least significant bit, the smallest number should be considered to be 0.5. In this case the dynamic range is given by,

$$20 \log \left[\frac{2^{15} - 1}{0.5} \right] = 96 \text{ dB}.$$

In general an N-bit fixed-point number system with a sign bit has a dynamic range of approximately $6N$ dB.

At first glance 96 dB seems like a generous dynamic range but it turns out that it is not enough for many tasks. Live music has a dynamic range of about 100 dB. Some fixed-point systems may go as high as 24 bits but even that may be insufficient for laboratory measurements.

Floating-Point Numbers

The use of a floating-point number system is the solution to the problem of extending dynamic range. In this number system the exponent controls the dynamic range.

For example, consider a 32-bit floating-point number with a 24-bit mantissa and a sign bit. The remaining 7 bits are available for the exponent. In this case the dynamic range is approximately $6 \times 2^7 = 768$ dB. In floating point numbers an N-bit exponent has approximately 6×2^N dB of dynamic range.

Example

Consider an audio system with a requirement for a 100 dB dynamic range. For a noiseless signal in fixed-point numbers this requires 17 bits plus one bit for sign. However, it is usually desirable to keep the noise level at least 60 dB below the signal level. The extra 60 dB requires another 10 bits, so the total requirement is 27 bits plus one bit for sign.

This same level can be achieved better in floating-point numbers because a 10-bit mantissa can provide the 60 dB of relative signal-to-noise and another 5 bits of exponent can provide a dynamic range of 192 dB. One more bit is needed for sign, for a total of 16 bits.

It is clear that a 16-bit floating-point system provides a better dynamic range than a 28-bit fixed-point system.

Typical specifications for a compact disc player:
Frequency response: 2 - 20,000 Hz ± 0.3 dB
Signal to noise ratio: >105 dB
Harmonic distortion: <0.004%
Sampling rate: 44.1 kHz
Number of bits: 16
Channel separation: >98 dB at 1 kHz.

Data Import

There are two operations that you will find useful in almost everything you do with spreadsheets in the real world. The first one is file transfer into spreadsheets from disk files. The second one is transfer from other applications such as data acquisition systems. Using *Windows'* Dynamic Data Exchange (DDE) turns your spreadsheet into an integral module in a laboratory or field test system.

First let's look at an example of data import by reading a disk file. In this example an A/D converter digitized a voltage and the result was saved to disk as an ASCII file. Use a file extension such as .TXT or .PRN for the ASCII file. Be sure to select this file type when you attempt to open the file, or just select "All Files (*.*)".

In *Excel* start with a new worksheet or a definite space in an existing worksheet and use the mouse pointer to select Open File (usually the second icon at the upper left-hand part of the screen). When you click on this icon with the left mouse button a Wizard will appear, as shown in Fig. A6-1.

Follow the directions and proceed to the next two parts of the Wizard. There is almost no way you can go wrong. If the ASCII file is delimited, *Excel* will detect this and take care of things. The result of this text import is shown in Fig. A6-2.

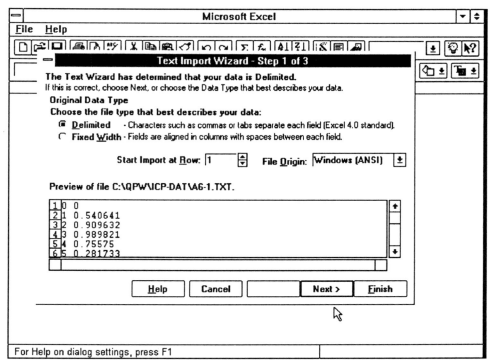

Fig. A6-1. Step 1 of the Text Import Wizard in *Excel.* If these choices are correct, proceed to the next two steps by clicking on "Next >" with left mouse button.

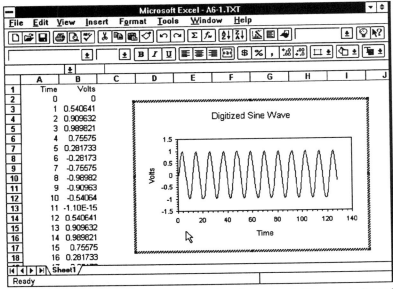

Fig. A6-2. ASCII file imported into *Excel* and graphed. The worksheet has 8192 cells in a column, so it is preferable to import data in column format.

In *Quattro Pro* click on Notebook, then Text Import and follow the directions. The text import box is shown in Fig. A6-3 and the result is shown in Fig. A6-4.

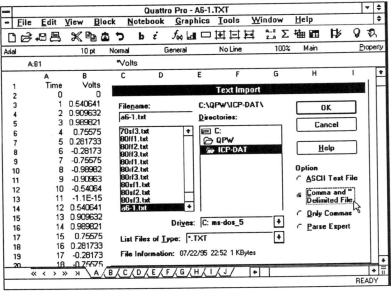

Fig. A6-3. Delimited file import in *Quattro Pro* 6.0. If you don't know the file format click on the Parse Expert button and consult on-screen Help.

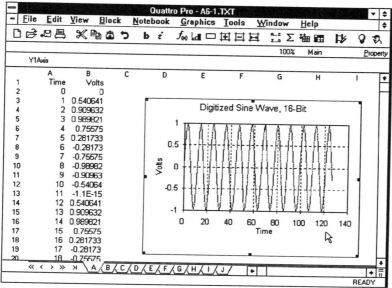

Fig. A6-4. ASCII file imported from disk into *Quattro Pro* 6.0 and graphed.

Figure A6-5 shows a 1-kHz sine wave digitized with 16 bits at a sampling rate of 44.1 kHz. Figure A6-6 shows the same sine wave with severe aliasing caused by under-sampling.

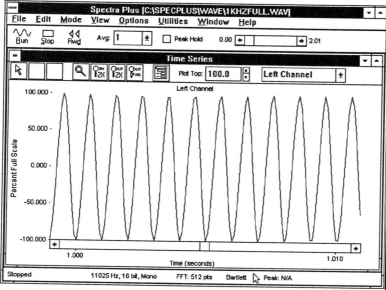

Fig. A6-5. Digitized sine wave sampled above the Nyquist frequency. This is the oscilloscope display in *Spectra Plus*.

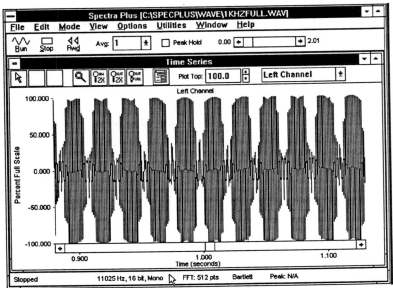

Fig. A6-6. Sine wave digitized at 16 bits, but under-sampled. Note severe aliasing.

Figure A6-7 shows a square wave with a dc component, digitized at 12 bits and 90 kHz sampling rate.

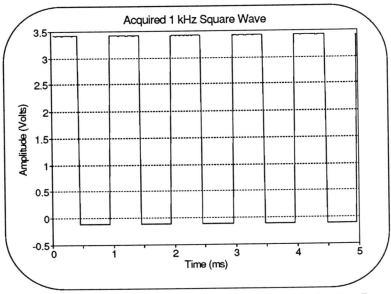

Fig. A6-7. This analog signal was digitized, imported into *Quattro Pro*, graphed, and exported as an encapuslated Postscript file for printing in this appendix.

Most data acquisition software will export ASCII or text files, which makes it easy to import these files into a spreadsheet for further data analysis and polishing into publication-quality graphics. For example, National Instruments' *LabView* has a "spreadsheet v.i." (virtual instrument) that you can use to put a virtual push-button on your screen to send acquired data to a spreadsheet.

You may have to select the proper parsing of the import file, depending on how the ASCII file is saved. Two common formats use either commas ("comma delimited") or tabs ("tab delimited") to separate data. Examine the ASCII file on your monitor before importing so you can identify the type of delimiting; you can open these files in a wordprocessor for viewing.

File import into Lotus *1-2-3* version 2.x is not difficult, but it's a bit more involved than *Windows* import. This will read data from a text file on disk into the worksheet beginning at the current cell. Move the cell pointer to a blank area of the worksheet, large enough to include the imported data. Type in the command /File Import. Then select one of the following:

- Text.   This imports each line of data as a long label and enters it in one cell of the worksheet. *1-2-3* enters each successive line of data in the same column below the cell pointer. Each line in the text file should end with a carriage return or a line feed and should not exceed 240 characters.

- Numbers.   This imports labels and numbers from a delimited file and enters them in separate cells in the worksheet, beginning in the current cell. When a text file is not delimited, Numbers imports only numbers and ignores text in the file.

Lotus *1-2-3* displays files with a .PRN extension in the current directory. To list files with a different extension, type the complete file name and extension, or type *.* and press Enter ⏎ .

Next, specify the name of the text file. *1-2-3* imports data from a text file either as text, numbers, or both, depending on the contents of the text file and the option (Text or Numbers) that you selected. *Note:* Numbers in a text file can contain decimal points, but numbers should not contain commas because commas act as delimiters.

Tip: If data are organized in columns in the text file but are imported as long labels, then use the /Data Parse command to place data into separate cells.

Dynamic Data Exchange (DDE)

This is a *Windows* communications protocol that allows sharing data among applications. You can configure *Excel,* for example, to send data and commands to other applications, and receive data and commands from other applications. Applications can act as both servers and clients, and DDE is not limited to *Microsoft* applications. DDE is one of the most powerful features of *Windows.*

Windows supports three type of interprocess communications, of which DDE is one. The second is manually-prompted data exchanges through the *Windows* clipboard, and the third is shared memory in a Dynamic Link Library (DLL). Consult your *Windows* manuals for more information.

You can use *Visual Basic* to develop your own DDE procedures. For example, you may want to read data from IEEE-488 bus instruments like digital multimeters and frequency counters, and move the data into *Excel* for automatic logging, analyzing, and graphing. There are only five commands to master: INITIATE and TERMINATE start and stop a DDE conversation, POKE sends data to another application, REQUEST acquires data from another application, and EXECUTE executes commands in another application.

A whole book could be written about DDE. For a quick over-view and an example of DDE with *Visual Basic* and *Excel* see Tolson's article (Digging Deeper, below).

Digging Deeper

Universal Laboratory Interface consists of a 12-bit data acquisition board and software. This is an inexpensive entry-level package that will connect your computer to the outside world. It is available from Vernier Software, 8565 S.W. Beaverton-Hillsdale Hwy., Portland, Oregon 97225. Phone (503) 297-5317. FAX (503) 297-1760.

The data acquisition board in the *Universal Laboratory Interface* is available separately from Sunset Laboratory, 2017 19th Avenue, Forest Grove, Oregon 97116. Phone (503) 357-5151. This is a half-size card that plugs into a slot in your computer and connects to external devices through either digital or analog voltages. The board has eight multiplexed inputs and three counter-timers for various operations, including producing output signals. A diskette with sample programs is included.

Principles of Digital Audio, K. C. Pohlmann (Sams, Indianapolis, Indiana, 1985). Also see *The Compact Disc*, K. C. Pohlmann (A-R Editions, Madison, Wisconsin, 1989). ISBN 0-89579-228-1.

Spectra Plus, an inexpensive package with top-of-the-line features, turns your computer into a digital audio lab. It works with most of the popular sound cards. This is a dual-channel digital oscilloscope and spectrum analyzer using the FFT. Other features include signal generation and the ability to produce a 3-dimensional plot to track time-changing frequencies (joint time-frequency analysis). Data can be exported to a spreadsheet. Pioneer Hill Software, 24460 Mason Road, Poulsbo, Washington, 98370. Phone (800) 401-3472.

A plethora of two-channel audio sound boards is available for personal computers. *Sound Blaster* compatibility is the standard. These boards usually include 16-bit digitizing at 44.1 kHz (a few go to 48 kHz), voice recognition, voice and music synthesis, audio compression and decompression, and audio editing. *Tip:* Choose a board that has the capability to send and receive simultaneously on both channels, like the Tahiti sound card produced by Turtle Beach Systems. This enables you to use your computer with *Spectra Plus* as a tracking generator for System Function measurements, Doppler applications, multiple cross-correlation, and similar procedures.

B. Tolson, "*Visual Basic*'s graphics, *Windows*' DDE turn standard applications into test-software modules," *Personal Engineering & Instrumentation News* **9**, 49 - 56 (1992), February.

Laboratory and industrial-grade data acquisition hardware and software are available from many sources. These include:

National Instruments, 6504 Bridge Point Parkway, Austin, Texas. 78730-5039, Phone (800) 433-3488.

LABTECH, 400 Research Drive, Wilmington, Massachusetts. 01887, Phone (800) 899-1612.

Keithley Metrabyte, 440 Myles Standish Boulevard, Taunton, Massachusetts 02780. Phone (800) 348-0033.

For more sources and updates on data acquisition products see recent issues of *Personal Engineering & Instrumentation News* (ISSN 0748-0016). Phone (603) 427-1377. FAX (603) 427-1388. Internet pgsperseng@aol.com.

Index